Jeremy C. Ganz

Gamma Knife Surgery

Second, revised edition

SpringerWienNewYork

Jeremy C. Ganz

M. A. (Cantab), Ph. D. (Bergen), F. R. C. S.
(London), Consultant Neurosurgeon
Honorary Professor of Neurosurgery, Department of Neurosurgery,
University School of Medicine, Graz, Austria

© 1993 and 1997 Springer-Verlag/Wien
Softcover reprint of the hardcover 2nd edition 1997

Data conversion: Bernhard Computertext, A-1030 Wien

Graphic design: Ecke Bonk

Printed on acid-free and chlorine-free bleached paper

Cover image: Fine beams crossing over at a focal region located at the site of a deep seated central target in the brain

With 49 Figures

Die Deutsche Bibliothek – CIP-Einheitsaufnahme

Ganz, Jeremy C.:
Gamma knife surgery / Jeremy C. Ganz. – 2., rev. ed. – Wien ; New York : Springer, 1997
ISBN-13:978-3-7091-7417-3

CIP data applied for

ISBN-13:978-3-7091-7417-3   e-ISBN-13:978-3-7091-6831-8
DOI: 10.1007/978-3-7091-6831-8

Lars Leksell (1907–1986)

This book is dedicated to three people. Firstly, with affection, to *my mother*, who first interested me in the study of medicine and then helped greatly with that study. Secondly, with respect, to one of the pioneers in the field of Gamma Knife surgery, *Erik Olof Backlund*, who led the author to the water of radiosurgery and persuaded him not so much to drink as to immerse himself therein. Finally, with admiration, to the late *Professor Lars Leksell*, without whom there would be no book to write. This edition is dedicated to my wife *Dr. Gao Nan Ping*, without whose constant support and encouragement there would have been no book.

Preface

This book attempts to combine many different threads into a comprehensible whole. Since the subject is the Gamma Knife and the author is a neurosurgeon, the field of clinical interest is restricted to intracranial pathology. The discipline of radiosurgery now applies to patients who may reasonably be referred by internists, neurologists, otolaryngologists, endocrinologists and several others. Some of the topics, touched upon, such as stereotaxy and the construction of a radiosurgical instrument are unfamiliar to the majority of medical men. Other topics, such as those pertaining to the reactions between radiation and living tissue, are not exactly unfamiliar and yet, for most of us, they are not comfortable areas of expertise: in that we have some basic knowledge but not enough to draw conclusions and interpret. In particular, it is not easy to answer the very sensible questions that patients ask, when being considered for this particular form of treatment.

The author has attempted to describe the basic relevant phenomenology in terms that should be readily understandable to a non-specialist physician. To do this, he has been heavily dependent on the expertise of a number of mathematically sophisticated collaborators, who have checked his manuscript. They are named in the acknowledgments section.

The relevance of the different sections of this book will naturally be assessed differently, according to the experience and interest of the reader. To simplify access to the information that is required, the book is divided into three main sections. Firstly there is an outline of the scientific principles underlying radiosurgery. This section may seem rather detailed, but to the best of the author's knowledge a basic acount of the various techniques and concepts has not previously been assembled in a single volume. Thus, some detail is necessary to ensure that the topics are presented logically. Of course, the reader can choose to avoid what seems either too detailed or too irrelevant for his needs. This section is concluded with a short account of how the most used radiosurgery machine, the Leksell Gamma Knife came into being and how it achieves its surprising and elegant accuracy. Secondly, there is a short section, consisting of a single chapter, describing the experience of Gamma Knife surgery from the patient's point of view. The patient's impressions and apparent requirements are outlined, based on what our patients themselves have told us. Thirdly, there is an account of current thinking on the commonest diseases treated in the Gamma Knife. This section is especially constructed to help a referring physician assess the appropriateness of the treatment for an individual patient. It also provides necessary facts relating to the patient's illness and this treatment method, as opposed to facts related to the treatment method itself. In the third section the author must again state his indebtedness, though this time to colleagues rather than collaborators. The reader will find a certain amount of repetition between the individual chapters. This is intentional to aid the reading of each chapter without being disturbed by cross reference. The text has drawn heavily on the extensive published material relating to the subject, particularly that from Stockholm and Pittsburgh. More detailed reading matter can be obtained by use of the reference lists attached at the end of most chapters. However, it is emphasised that, while data have been obtained from these various expert sources, the opinions and any inaccuracies to be found in the text are the responsibility of the author alone.

At the end of the book there is a short concluding chapter, indicating directions of future interest and expansion, for what seems to be one of the most exciting new therapeutic technologies of recent years.

In conclusion, the author would like to draw the reader's attention to an often forgotten aspect of the use of ionizing radiation in the treatment of disease. *The results and complications of treatment using radiation will be related to the* **dose.** Physicians, who are themselves extremely careful in the dose of pharmacological agents seem to forget that dose is also important in respect of the effects of radiation treatment. Thus, the first question that must be asked when comparing results from one treatment with another is – *what was the dose?* If this simple information is not available then assessment of the results is impossible. There are of course other variables which may be considered but dose is essential. The author emphasises this point because practice with respect to dose has changed over the years, as have the results of treatments using radiation. Intelligent assessment of the results of treatment thus requires awareness of the dose used and this book is intended to facilitate intelligent assessment of the patients to whom it applies.

Jeremy C. Ganz

Preface to the Second Edition

During the time following publication of the first edition, radiosurgery has become widespread. This implies a necessity of emphasizing the basic principles on which the method is based. As more and more patients are treated by radiosurgery the stochastic chance of error in application of the method increases. In this context the words of the Czech writer Milan Kundera bear repeating. "A worker may be the hammer's master, but the hammer still prevails. A tool knows exactly how it is meant to be handled, while the user of the tool can only have an approximate idea". The intention of this book remains to make available in a hopefully comprehensible way the basis of knowledge on which Gamma Knife surgery may be safely based. If this aim is fulfilled then the chance that the technology will be used in an appropriate way should be increased. It remains the intention that the book should be of assistance both to those who are starting to use the Gamma Knife and to whose who wish to refer patients for treatment.

The body of the text from the original edition has been largely unchanged. Those errata which have been discovered have been corrected. Nonetheless, there are changes. There has of course been progress in what is a rapidly developing field. To incorporate newer knowledge an appendix has been added which attempts to outline newer information and changing opinions. This appendix is divided up into sections related to the different chapters so that the changes may easily be related to the original text. In this way, it is hoped that the reader may more readily gain an impression of the developments that have taken place, by comparing what is new with what is not so new.

An extensive reference list (423 references) has been added at the end of the book for those who wish to delve deeper into the subject. The vast majority of these references relate only to the Gamma Knife. Exceptions are articles concerning the scientific basis of radiosurgery, irrespective of the technique involved. Also clinical papers of significance from other methods are included if their contents contain information of which the Gamma Knife user should be aware.

Finally, the author would like to thank those who have shown such interest in the first edition of this book that it has been possible to publish a second one.

Jeremy C. Ganz   April 1997

Acknowledgements

Nobody can write a book without help from other people. The first step is to obtain relevant background literature and the staff of the University Library in Bergen have been unflagging in their efforts to acquire references promptly. The next step is the writing itself, a time-consuming grind made immeasurably easier by the development of personal computers and their accompanying software. This book has been written on a notebook PC made by the English company Reeves plc of Northampton, and it has been a joy to use. Most of the text has been written in Microsoft Word and set in Xerox's Ventura Publishing. The latest additions have been written in Microsoft Word for Windows. Without these technical aids I doubt that it would have been possible for the author, a busy clinician, to write a document of this length. Not least, the compilation of an index was not the torture of frustration described by older colleagues; merely a protracted annoyance. Hand in hand with the writing goes the production of illustrations. Not for the first time, I should like to express my deepest thanks to the University Department of Photography, for their cheerful enthusiasm and helpfulness.

The contents of this work cover a wide variety of fields and it has been natural to request various experts to check the manuscript for factual accuracy. It must be said that the generosity of these experts with their time and patience has been impressive. Jörgen Arndt esquire, of Elekta Instruments AB advised on the chapter on radiophysics. He has been involved with the Gamma Knife from the earliest days and when he is asked to help no trouble seems too great. I should also like to thank Professor Herbert van der Kogel of Niejmegen in Holland. He is a leading authority on radiobiology, especially the radiobiology of the CNS. Despite his punishing schedule, he made the time to analyse the relevant chapters thoroughly and had many invaluable comments and corrections. The introductory chapter, the chapter on stereotactic technique and the chapter on diseases related to the pituitary region were read through by Professor Erik-Olof Backlund, my chief, whose interest and expertise in these areas is well known. Dr. Christer Lindquist read the chapter on arteriovenous malformations and made a number of interesting comments. Dr. Georg Norén has given the author much advice about the treatment of acoustic schwannomas, over the years. He checked the chapter on this topic for factual accuracy. The chapter on the development of the Gamma Knife was checked by Dr. Dan Leksell. His help was also prompt and generous, not least in reviewing historical material. Professor Backlund was also most helpful with the history of the early days of Gamma Knife treatment. He has extensive documentation of the earliest operations and he willingly made this available to the author. I should like to thank my brother Robin Ganz esquire, an English scholar, who kindly checked that the language was not too obscure.

In the daily work with the Gamma Knife many people are involved. If there were no patients there would be no basis for this book. In this respect I should like to mention two of our Nursing Sisters, Anne-Lise Røssland and Inga Sekse, who are responsible for all the work involved in arranging appointments and admissions and who are never less than helpful. I must also thank the radiographers Aud Bruland and Britt Tønnebekk, who supervise the CT and angiography services. They are overworked but never let this come in the way of doing their best to keep the doors of the X-ray department open for Gamma Knife patients. Also, I have to thank nursing sisters Kari-Britt Overå Mathisen and Marga Larsen, of the operating room at Haukeland Hospital, and their staff. They unfailingly deliver the stereotactic frame and all its accessories to the various locations outside the operating theatre, where frame application is undertaken, thus facilitating an effective service. Moreover, I am grateful to the many patients who have been through our hands, since we started in 1988. It has been striking how much they like the treatment and it has been a pleasure and privilege to get to know them.

Many colleagues have contributed to the establishment of Gamma Knife treatment in Bergen. Professor Backlund acquired the machine and is the author's mentor in its use. This debt cannot easily be repaid. Particularly valuable is his teaching, that no measurement or adjustment may be made by a single person. Everything must be checked by a third party. The radiologists, Professor John-Ludwig Larsen and Drs. Jostein Kråkenes, Alf Inge Smievoll and Gunnar Moen, have devoted much time and patience to the pre- and post-treatment assessment of patients and of course in helping to ensure that the targets are defined with maximum accuracy. Drs. Juan Robbie Mathisen and Paul-Henning Pedersen, who are the other members of the Gamma Knife team in Bergen have stimulated the author constantly, during treatment and in many discussions. Also I should like to thank all those colleagues who refer patients. It has been interesting to see how initial uncertainty as to the method's validity changes to confidence. Thus, once a colleague has established contact with the Gamma Knife by sending one patient, most often that colleague will send other patients later.

Nobody can use the Gamma Knife method without the assistance of radiophysicists. The head of the Department of Radiophysics at Haukeland Hospital, Anfinn Mehus has been most helpful in the establishment and daily running of the Gamma Knife. At the start Harald Valen and subsequently Frits Thorsen have given invaluable assistance during dose planning. In particular the latter, for whom the Gamma Knife is now his major field of work, has been consistently available, innovative and enthusiastic in attempting to improve the technical quality of the treatments that we give. I have to say that working with these two physicists and indeed contact with their entire department has been nothing but a pleasure.

Finally, I should like to express my gratitude to the Gamma Knife itself and to its inventor Professor Lars Leksell. This machine is a constant source of satisfaction, with its elegant simplicity of concept and its equally elegant simplicity in daily use. It is undeniable that a user must have acquired some fairly extensive knowledge, prior to using the machine. In particular neurosurgical experience of three-dimensional anatomy, at open surgery is a pre-requisite for rational dose planning. However, given such knowledge, while individual targets may present a challenge to the dose planner, in principle, the Gamma Knife is remarkably easy to use.

*

It is of course very gratifying for the author that this book has been so well received and that a second edition is possible. I should like to acknowledge a particular debt to Professor Gerhard Pendl, the chief of neurosurgery at Karl-Franzens-Universität, Graz, Austria. He has been most generous to the author in many ways and has given much helpful advice relating to radiosurgery. It is also a pleasure to thank Springer-Verlag for their work with this book. They have been consistently helpful and always most efficient in all matters related to publication.

Table of Contents

I. Introduction and Basic Principles

II. The Patient's Experience

III. Clinical Aspects

I. Introduction and Basic Principles

1. Introduction

The impulse to perform surgery antedates literacy by some thousands of years, as demonstrated by the large number of prehistoric trephined skulls that have been discovered. These earliest operations were based at best on superficial empirical experience and not on knowledge, in any way in which that term could be understood today. Thus it is remarkable that they were performed at all. What is even more remarkable is that the "patients" allowed themselves to be subjected to such surgery. All this suggests that the need to operate or to suffer surgery is primitive and not entirely rational. In fairness to the stone age surgeons, many of these prehistoric trephine openings show signs of healing. Thus their operations were in fact often successful, measured by the yardstick of technical success.

The next stage in neurosurgical development following prehistoric trepanation came with Hippocrates. This remarkable man told us "First do no harm". Even so, the same Hippocrates, with very little if any neuroanatomical knowledge described in detail the indications and procedures to be followed in the treatment of skull injuries. His practical advice, for example on cooling the trephine was excellent. His indications for surgery were often, by any rational modern standard, bizarre.

These few examples could be multiplied ad infinitum. The point is that even wise people can be tempted to perform surgery in situations where the indications are dubious. And it must be admitted, there is something theatrical and attractive about the whole concept of a surgical operation. There is the dramatic setting, in an operating room, with all its attendant technology, ritual and regalia. There is the dramatic procedure, with its attendant risks and rewards. However, it is an undeniable fact that all surgery is potentially dangerous. There cannot be a surgeon alive who has not, when reoperating a patient, witnessed the scar tissue resulting from previous surgery, itself a witness to the fact that all surgery exacts a price. Happily, such scar tissue usually does no apparent harm but its presence is a reminder that the body resents mechanical interference. In addition, all operations carry with them local risks of bleeding and infection in the operating field and more systemic risks related to immobilization.

Thus, it is perhaps wise to consider the advice of earlier generations and accept surgery as a treatment which is appropriate when all else has failed. "Desperate diseases require desperate remedies".

None of the foregoing is meant in any way to minimise the achievements of modern surgery. The author is himself an active neurosurgeon, who enjoys and is convinced of the value of his profession. Nonetheless, while good surgery is always acceptable and where used, most often unavoidable, it is a brave person who would describe it as desirable. It is more in the nature of an elegant necessary evil. This notion gains support from the knowledge that while people have allowed themselves to be subjected to strange surgical procedures, from prehistoric times, there is nonetheless a tradition in literature, indicating a yearning for more painless methods of treatment. Thus, Homer describes the miraculous effects of ointments on wounds in the Iliad and Christ heals by strength of spirit in the New Testament. And while such treatments are still, in the main, unattainable, it must be sensible to long for them. Anyone can understand a patient's reluctance to undergo surgery, with the resulting pain and possibly disfiguring scar. Concern about disfigurement is particularly worrying for the neurosurgical patient. The consequences of even successful neurosurgical procedures can be very distressing for a particular reason. Anyone with a dent in the forehead, or with epilepsy or a paralysis is viewed by many people as having a changed, disturbed or indeed deranged personality. This is a colossal burden to bear and thus to be avoided at all costs, though it may of course be an unavoidable price to pay for staying alive.

Bearing the above in mind, the surgeon takes precautions against the dreaded complications of wound haemorrhage and wound sepsis, by applying meticulous asepsis in the operating theatre and by operating with a gentle technique which respects the living tissue under the knife. He also relies heavily on the body's ability to compensate for the surgical invasion. For neurosurgeons it is particularly important to gain as much space as possible within the crowded intracranial cavity, thus reducing the force necessary to retract the brain. The brain may react badly to prolonged retraction, particularly if too great a force is applied. To this end, the neurosurgeon releases CSF before applying the retractor and the neuroanaesthetist manipulates the intracranial circulation to minimise blood volume, without prejudicing tissue perfusion. Moreover, the neurosurgeon knows just how surprisingly large quantities of the brain may be removed, without causing social embarrassment to the patient.

The principles of gentle sterile surgery based on knowledge of and respect for the body's compensation mechanisms is part of the mainstream tradition of medical development in this century. As medical men tend to be conservative, it would take a powerful and courageous intellect to break with such a tradition. Lars Leksell, for many years Professor of Neurosurgery, at the Karolinska Hospital in Stockholm was possessed of such an intellect. He was certainly one of the most creative neurosurgeons of this century. Over a period from the 1940s to the 1980s, he devoted his time to methods of treatment which were not confined to taking advantage of the compensation mechanisms that make open surgery possible. On the contrary, his work seems to have had a central aim to reduce operation trauma to an absolute minimum. In his monograph, *Stereotaxis and Radiosurgery*, he states: "The tools used by the surgeon must be adapted to the task, and where the human brain is concerned they cannot be too refined".

One of Leksell's first clinical contributions was to devise a stereotactic frame for routine use in humans. Prior to this, stereotaxy had been primarily an experimental tool, though a stereotactic technique had been used in the treatment of trigeminal neuralgia and for intracerebral targets. The advantage of the Leksell system was that it was relatively simple and versatile in operation, when compared with other contemporary stereotactic systems. As a result stereotactic surgery gained an impetus which has been maintained to the present day. With Leksell's system, access could be gained to any intracranial region with minimal trauma. However, Leksell's attempt to minimise operative trauma did not stop with the design and further development of a clinical stereotactic system. He went further and with a small group of associates devised an apparatus for treating intracranial pathological processes, without opening the cranium. This instrument, the Leksell Gamma Knife, was designed for use with the Leksell stereotactic system.

The terms radiosurgery and Gamma Knife have been the source of some controversy. Those who use radiosurgical techniques would justify the use of the term as follows. When ionising radiation is employed to damage or destroy a pathological process, it is vital that normal tissue in the neighbourhood of the lesion remains undamaged. This is achieved in conventional radiotherapy by fractionating the dose and by directing the radiations first from one side and then from the other. The beams are few, broad and seldom more than 8 different beam directions are used. With the Gamma Knife technique, there are over 200 beams of radiation and they are individually very narrow. This arrangement enables the construction of a very precise radiation field, limited to the pathological lesion. Normal tissue is excluded from dangerous levels of irradiation, because in radiosurgery a correlate of the very precise radiation field is a rapid fall in radiation levels just beyond the edge of the lesion. It is the surgical precision of the radiation field, administered at a single session, that has led to the term radiosurgery.

The Gamma Knife is the first radiosurgery instrument which has gained widespread use in clinical medicine. Its design will be outlined in a subsequent chapter. However, it in no way resembles any ordinary knife. It is a massive machine which weighs about eighteen tons. Nonetheless, it delivers an exquisitely exact field of radiation. Thus if the technique is called radiosurgery the instrument performing the treatment is by analogy a radiosurgery knife. Since the radiation source is ^{60}Co which emits gamma radiation, this particular radiosurgery instrument is called a Gamma Knife.

Subsequent chapters will outline the development and application of the Gamma Knife technique. Its strengths and weaknesses will be described. Future trends will be discussed. However, wherever the future may lead, the existence of this technique, which enables neurosurgical treatment to be performed without opening the skull, is very much the result of one man's vision and his insistence on reducing trauma not by just relying on physiological compensation, but by being as nearly as possible, truly atraumatic.

Suggested Further Reading

1. Leksell L (1949) A stereotaxic apparatus for intracerebral surgery. Acta Chir Scand 99: 229–233
2. Leksell L (1971) Stereotaxis and radiosurgery. An operative system. Ch C Thomas, Springfield IL

2. Principles of Stereotaxy

Introduction

There are few parts of the body which may not be approached by means of a surgical operation. An exception to this rule is the deep parts of the brain. The reasons are easy to understand. The head is a rigid cavity and about 90 % by volume of its contents is cerebral tissue. This tissue tolerates retraction and displacement badly. Moreover, access to, for example, the basal ganglia involves an unavoidable and unacceptable risk of damage to eloquent areas lying superficial to or adjacent to these ganglia. Thus, conventional open surgical techniques are an essentially inappropriate method to use for operations in the depths of the brain, irrespective of the sophistication of modern microsurgical techniques. These problems have been appreciated for a long time. A solution, at least in principle, has also been around for a long time, in fact since before the first world war.

The famous British neurosurgeon and neurophysiologist, Victor Horsley, wished to gain access to deep cerebellar structures in animal experiments. He sought the help of an Oxford mathematician and medical man, Robert Clarke and together they published an account of the first stereotactic instrument in 1908. However, neither this instrument, nor any modification of it were ever used in clinical work, though one was designed but abandoned for lack of interest. Approximately forty years were to pass before two Americans, Spiegel and Wycis in 1947 published the results of their operations with a stereoencephalatome, as they called it, for pain and psychiatric disorders.

The term stereotaxy derives from two Greek roots "stereos" meaning solid and "takse" meaning arrangement. However, in the past there has been an at times passionate debate, as to whether the correct adjective from stereotaxy should be stereotaxic or stereotactic. The former is etymologically correct. However, the latter was felt to be somehow more in keeping with a surgical procedure, in that "tactic" is derived from touch. But! It is derived from the *Latin* word for touch, so that etymologically speaking stereotactic is a chimera. Nonetheless, in 1973 the international body responsible for furthering the interests of those involved in this form of surgery changed its previously rather cumbersome name to "World Society for Stereo**tactic** and Functional Neurosurgery". Thus, this is the form which will be used throughout this book.

Principles

While knowing what the word stereotaxy means may indicate the area of interest, it does not tell us any more about the principles of the method. The aim of the technique is to relate the location of deep and inaccessible intracerebral structures to a three dimensional Cartesian axis system. The first step in this process is to enclose the head in such a system. This is done by fixing a rigid metal frame to the head. The borders of the frame then constitute the Cartesian axes, while the cranium serves as a platform to support the frame and the cerebrum is enclosed both physically and conceptually within a microcosm, where

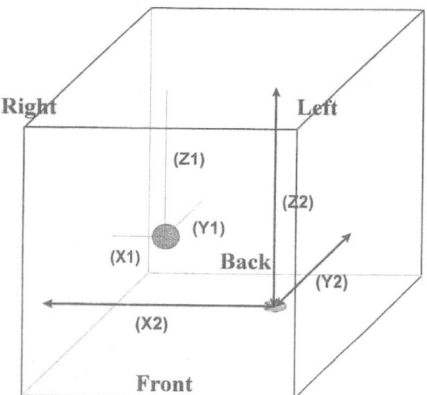

Fig. 2.1 Stereotactic Principle (I)
Stereotactic technique relates the position of intracranial targets to visible cranial or extracranial markers. The markers used today consist of a frame which is also a Cartesian axis system. The targets are related to the sides of the frame by perpendiculars dropped from the frame to the target. As can be seen, only one point is identified by the values $X1$, $Y1$, $Z1$ just as only one other and quite distinct point is identified by the values $X2$, $Y2$, $Z2$. Thus any intracranial location can easily be identified in relation to the frame which is fixed to the head. The orientation of the frame is a secondary consideration. It is the frame and not the head which is used to localise. This is useful in Gamma Knife work where it may be convenient to place the frame eccentrically or to rotate it in respect of the head

every point can be precisely defined in space. The way in which a point within the frame is defined in terms of the three Cartesian axes is depicted in Fig. 2.1 (page 8). Thus, it is possible to define any intra-cranial target in respect of the frame. However, to be of any use the frame must itself be a platform for a device which will hold an instrument or electrode, to be introduced into the brain, to reach the target. The small diameter of the instruments used and the mechanical stability with which they are held and introduced, by means of a rigid holder and guide, and the extreme accuracy of the localization implicit in the method are the basis of the exceptionally atraumatic nature of stereotactic procedures.

The Leksell System

A great variety of different stereotactic systems have been designed over the last forty years. Each system has its protagonists and its special fields of application. For a description of the essential technical principles of stereotactic surgery reference will be made to the Leksell system, because it is the system used in Gamma Knife surgery. These principles are on the whole independent of the system used, though the way in which technical problems are solved differ from system to system. As indicated in chapter 1, while the Leksell stereotactic system was not the first, it was the first to be constructed with a view to ease of application and frequent use. The sides of the cubic frame constitute the axes of the instrument. An arc is mounted on the instrument in an adjustable holder which is regulated in respect of the desired values in the three axes. The arc may be rotated backwards and forwards, with respect to the frame. The instrument holder is mounted on the arc and may be moved transversely across the whole circumference of the arc. When the axis values for the target point have been determined, the centre of the arc will always coincide with this target point. This arrangement allows a needle to be pointed at its target from an almost infinite number of directions. Thus, it is simple to design an optimal trajectory for the instrument to be introduced into the depths of the brain, avoiding especially sensitive structures, for example eloquent brain or important blood vessels. An important point of the design is that, for a given target setting, the point of the instrument introduced to the centre of the arc does not move. It does not move irrespective of how the direction of the shaft of the instrument is varied, by moving the arc backwards and forwards or by moving the needle holder transversely across the arc. This effect is quite uncanny and is illustrated in Fig. 2.2.

Fig. 2.2 Stereotactic Principle (2)
This is an illustration of a stereotactic frame, with a needle mounted, as it would be for penetration to a specified intracranial target. The picture is retaken with the needle approaching the target from a variety of directions. As can be clearly seen, the direction of the needle in no way affects the positioning of the point, which is placed at the desired target, as determined by appropriate adjustment of the various axes of the system

The names of the axes are of course X-axis, Y-Axis and Z-axis. However, the direction of the disparate axes varies from stereotactic system to stereotactic system. The choice of which axis points in which direction was originally arbitrary. However, with the Leksell system, as with most others in use today, the convention has become that the X-axis runs from side to side, the Y-axis runs from behind forwards and the Z-axis runs from above downwards. For anyone who wishes to remember the direction of the axes, the common zero point is above and behind the right ear.

Target Identification

The features of a stereotactic system, which have been outlined so far describe how a point in space, a so-called target point may be defined in terms of the reference axis system, built into the stereotactic frame, that is fixed to the head. If the target is "visible", for example a space occupying lesion, then a general knowledge of cerebral anatomy together with adequate imaging techniques will suffice. However, in the early days, stereotactic technique was almost exclusively used for the treatment of functional disorders. The targets in this situation are "invisible", consisting of discrete nuclei or tracts within larger anatomical entities, such as the thalamus or the basal ganglia. To locate such targets, a map of the region is required or rather a collection of maps in an atlas. An atlas of the internal cerebral anatomy of a variety of laboratory animals had already been produced by Horsley and Clarke, in the first decade of this century. The production of a human stereotactic atlas in 1952 was one of the major contributions of the pioneers Spiegel and Wycis, mentioned above.

While an atlas will show where a given nucleus or tract should be it will not show its precise location in an individual patient. It is necessary to have some visible structure with a constant relationship to intracerebral structures. In Horsley and Clarke's experimental work, skull landmarks were used. They had no choice in this, because contrast materials, which could outline soft tissues on X-ray were not available, at the time they performed their studies. They were aware of the limitations of this method of localization and checked the location of the lesions they made at post-mortem. Their physiological studies included only those animals where the lesion was correctly placed. This sort of inaccuracy would clearly be inappropriate for operations performed in human patients, so some other means of localization was necessary. To accomplish this, an internal cerebral reference system had to be defined. Air ventriculography had been first described by Harvey Cushing's pupil, Walter Dandy, in 1919. This technique was thus well established at the time when Spiegel and Wycis performed the first stereotactic operations, in humans. They outlined the third ventricle and used a line drawn between the posterior border of the Foramen of Monro and the anterior border of the pineal body, as a reference. They could then relate the anatomical structures, as depicted in their atlas, to this line. Moreover, they used the technique to demonstrate just how great the variability of the relationship between intracerebral structures and skull landmarks really is. Subsequently, Tailairach in France used a line joining the anterior and posterior commissures, visualized on an air ventriculogram of the third ventricle, as a reference in his atlas. This reference line is more constant than that based on the Foramen of Monro and the Pineal Body. This is because the borders of the former can be difficult to see and the latter varies a lot, in terms of size and ease of definition. Since that time the <u>intercommissural line</u> has been a standard reference in functional stereotactic surgery. The adequate definition of the posterior commissure required a somewhat refined air ventriculography technique. The subsequent introduction of iodinated positive contrast substances, in particular the discovery of water soluble contrast media greatly improved the quality of the anatomical definition of the boundaries of the third ventricle. This technique is still used in many centres during treatment, because it is still unusual to find a CT scanner in the operating room. Moreover, repeated ventriculograms can be used not only to identify a target location but also to check that the instruments are placed correctly at the target. Thus, with adequate X-ray definition and using the intercommissural line and an appropriate stereotactic atlas, the location of "invisible" targets became reliable enough for systematic clinical use. The "invisible" targets had been made "visible".

Newer Localization Methods

The treatment of pain, Parkinsonism and psychiatric disturbance, which required the localization of invisible targets remained the dominating indication for stereotactic surgery for approximately twenty years. However, stereotactic technique could also enable completely accurate placement of instruments into deep-sitting space occupying lesions, where this is appropriate. And it is clearly desirable to obtain a biopsy from those deep-sitting lesions that are tumours and puncture those that are cysts, haematomas or abscesses. Prior to the stereotactic era, these procedures were performed free hand, with a considerable margin of error. It is interesting, in this context, that Leksell's first stereotactic operation on a patient, using his own instrument, was to instill radioactive isotopes into a cranio-pharyngioma cyst. The problem in the early days, with using stereotaxy in the investigation and treatment of space-occupying lesions was that the X-rays available were inadequate for localization in many cases. Angiography could help with those lesions which show pathological vessels but often distortion or displacement of blood vessels is all that can be seen and this is too imprecise. Even so, with the advent of radiosurgery, the angiogram remains the examination of choice in the treatment of arteriovenous malformations. The various techniques for displaying the CSF spaces are really only adequate for the stereotactic localization of lesions which distort these spaces. Thus, identification of smaller lesions, deep in the cerebral parenchyma was difficult. However, the development of computer assisted tomography (CT) and more recently magnetic resonance imaging (MRI) have greatly facilitated stereotactic procedures. These techniques render space occupying lesions truly visible and thus simple to localize. Moreover, the modern machines can incorporate software which will superimpose a stereotactic grid on the film and this makes localization extremely simple (Figs 2.3–2.8). This use of the stereotactic frame, with CT and MRI units necessitated the development of an adaptor to fix the stereotactic frame to the CT/MRI table, to ensure that the axis system of the frame and the axis system of the CT/MRI software were concordant. A word of caution must be mentioned in respect of using MRI for target localization. There is a degree of anatomical distortion in MRI pictures, which increases from the centre to the edges. This distortion must be minimised by repeated checks. It makes central placement of the target within the frame even more important than otherwise; a topic to which we shall return.

The use of CT or MRI equipment virtually eliminates one technical problem, which has so far not been mentioned. This is the orientation of the X-rays to the film and the magnification of the target. This orientation had been a matter of great concern to the surgeon using stereotactic techniques, from the earliest days. This is because the precise definition of a target, with respect to the frame assumes that the central beam of the radiation, used to make the pictures is at a precise right angle, both to the frame and the film. If this is not so, then an unquantifiable element of distortion and thus inaccuracy creeps in, which is incompatible with the precision of stereotactic technique. For the Gamma Knife surgeon, controlling the orientation of the X-ray beam to the frame is most often relevant during angiography, used in the treatment of arteriovenous malformations. The degree of magnification is calculated from the films. The scales are seen on X-ray as a larger and a smaller ruler, the larger being nearer the X-ray tube and the smaller being further away (Fig. 2.9). It is relatively simple to calculate magnification from the difference in the magnification of the two scales. However, it should be repeated that such calculations only give an accurate answer if the central X-ray beam is at a right angle to the stereotactic frame. With computer assisted imaging techniques, the frame's position can be adjusted with reference to a built-in laser marker, thus ensuring that the various stereotactic axes are correctly aligned. Indicator plates, which are visible on the CT or MR are then fixed to the frame. As the indicators are correctly orientated, at a right angle to the plane of the films obtained, it is a simple matter to define the centre of the frame, using the indicator plates. Thereafter, a grid, centred at the centre point of the frame is superimposed. These manoeuvres make localization of a target, in respect of the axes of the frame, even easier, in that the target coordinates can be read directly from the film.

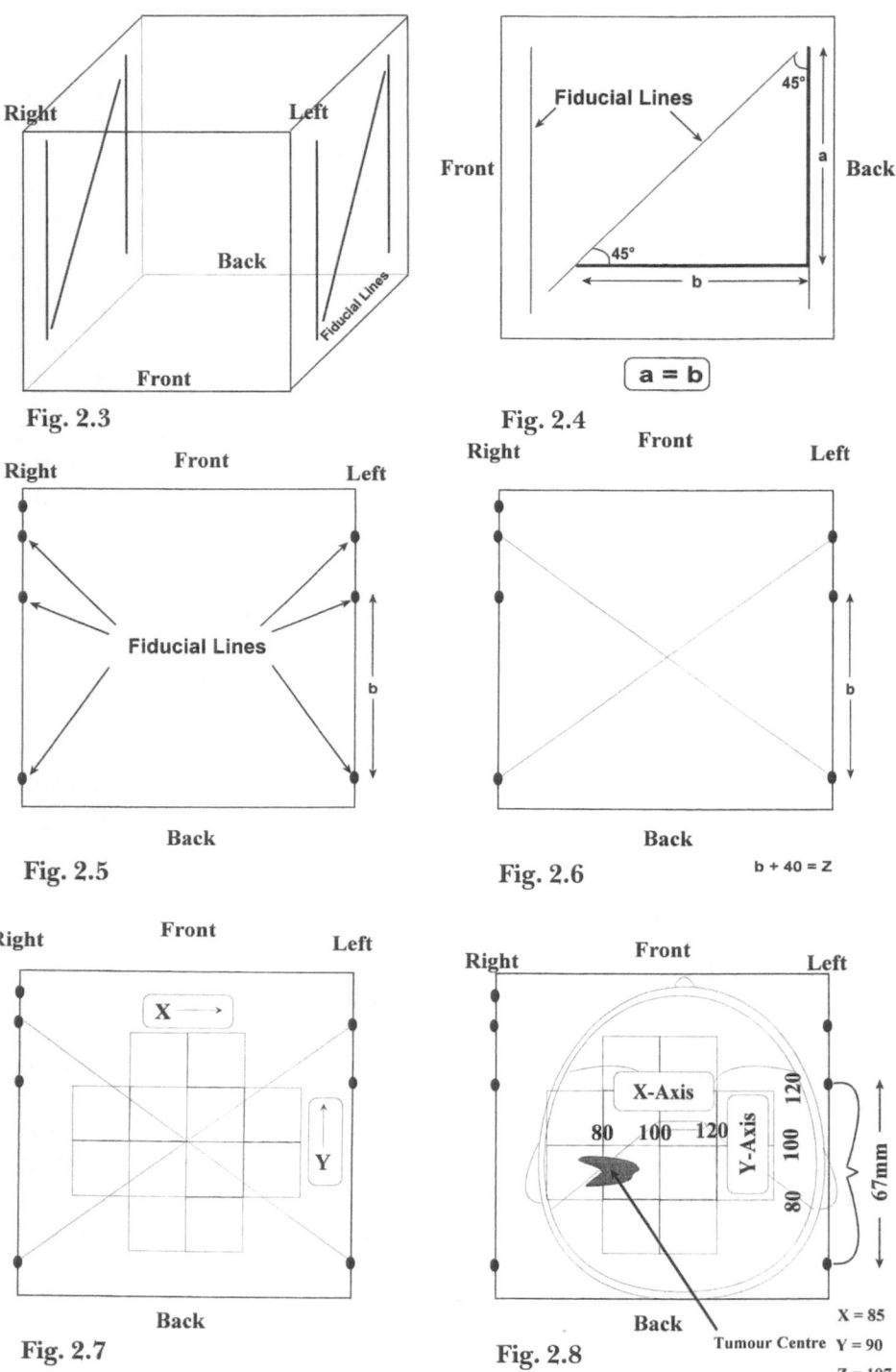

Fig. 2.3

Fig. 2.4

a = b

Fig. 2.5

Fig. 2.6   b + 40 = Z

Fig. 2.7

Fig. 2.8   Tumour Centre

X = 85

Y = 90

Z = 107

Modern Indications for Stereotaxy

Stereotactic technique, following the advent of CT imaging has a large number of indications, most of which have been mentioned. There has however been a tendency for its use to be restricted to a relatively small number of enthusiasts. It is not, even today, used routinely by all neurosurgeons. This is partly because, in the majority of centres, it has been used for functional work, which requires a specialised neuro-physiological knowledge, that is not a part of all general neurosurgical training programs. Furthermore, the basis of the technique is not technical surgical virtuosity but rather the avoidance of the need for such virtuosity. The Karolinska Hospital Neurosurgery Department, under Leksell's aegis taught that *stereotaxy was not to be considered an alternative to other forms of treatment but to be used in addition to the more*

Figs. 2.3–2.8 CT/MRI Indicator

Fig. 2.3: For stereotactic technique to work, the frame and target must be clearly visible on X-ray pictures. With CT or MRI imaging this is done by means of an indicator, into the walls of which X-ray opaque lines are fixed. These are called fiducial lines. This illustration shows a diagram of such an indicator. The lines are in the form of a backwards "N" with an extra vertical line at the front on one side to indicate the orientation. The diagram shows the lines and their orientation. **Fig. 2.4:** Left side of Fig. 2.1 seen in a profile. Because the oblique fiducial line is drawn at a 45° angle to the vertical fiducial lines, $a = b$. This is because a right angled triangle one of whose other angles is 45°, as in this case is automatically an isosceles triangle. Because $a = b$ it is possible to assess the "Z" value for a given CT slice, by measuring the distance b. **Fig. 2.5:** This illustration shows how the fiducial lines, shown in Figs. 2.1 and 2.2 appear on a CT/MRI slice as dots. It also illustrates where the distance b is measured to determine the "Z" value for the slice. **Fig. 2.6:** This shows how the indicators are used to determine the centre of the frame and the "Z" value. The frame centre is found by drawing diagonals from the opposite vertical fiducial lines to each other. Where these lines cross is the centre of the frame. The "X" and "Y" values at this point is 100 for both axes. The b distance can be measured from the CT/MRI images as seen and the "Z" value determined. "Z" = "b" + 40 because the posterior vertical fiducial line is at 40 in the "Y" axis. **Fig. 2.7:** This shows the superimposition of the grid concordant with the centre of the system, as determined in Fig. 2.3. The directions of the "X" and the "Y" axes are indicated. **Fig. 2.8:** This shows a diagram of a CT slice, with an acoustic neurinoma, with the target values for a potential gamma knife "shot", that is one dose of a multi-dose dose plan. Note that the "X" and "Y" values are less than 100, because the lesion is behind the midpoint of the frame and towards the right. The "Z" value for this CT slice = b + 40 mm, as explained in Fig. 2.6

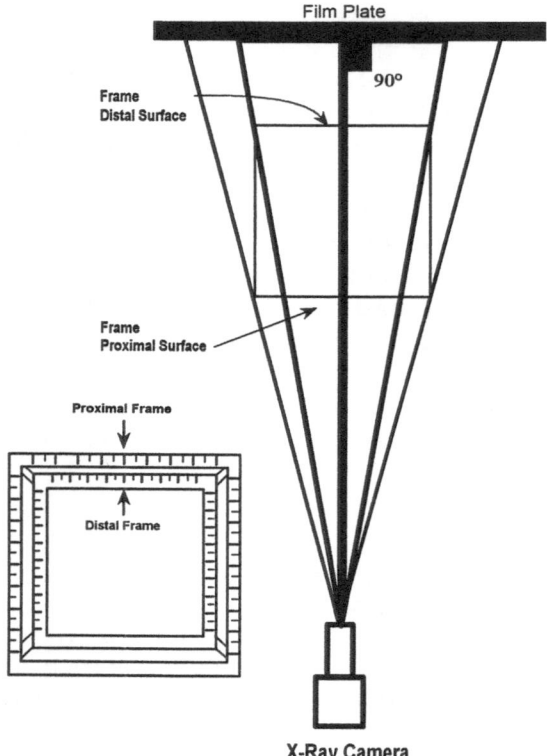

Film Plate

90°

Frame
Distal Surface

Frame
Proximal Surface

Proximal Frame

Distal Frame

X-Ray Camera

Fig. 2.9 Angiogram Indicator
When conventional X-rays are used, for example with angiograms for
arteriovenous malformations, it is vital that the central beam of the X-ray is
at a right angle to the X-ray film plate. If it is not, then there will be a
completely unknowable source of error in the geometry of the images taken,
making them useless. It can also be seen that the rays striking the proximal
side of the frame will project a larger image on the X-ray film than the distal
side. Thus, during assessment of the target location, the smaller image is
always furthest away from the X-ray camera and closest to the X-ray film plate

traditional armamentarium of neurosurgical options. Thus, this department, more than most others has for the past twenty-five years
concentrated on applying stereotactic methods to a great variety of
indications. Moreover, all neurosurgical trainees at this department
are expected to master the stereotactic technique early during their
training. The rationale is clear; as stated in the opening chapter. Any
method which reduces the per-operative trauma of a neurosurgical
procedure to a minimum must be considered as preferable to one

relying on surgical virtuosity, which takes advantage of the brain's compensation mechanisms. There are then three main areas of indication, which have been used for open stereotactic procedures. Firstly, there is the original group of functional illnesses already mentioned. Then, there are the direct primary treatments of cystic space occupying lesions, for example the installation of radioactive isotopes into craniopharyngioma cysts and the aspiration of abscesses. Moreover, aspiration of intracerebral haematomas may also be performed stereotactically, though the timing and indications remain a matter for debate. Finally, solid tumours may be biopsied as a preliminary to rational treatment planning. Not only that, but where indicated, a guide – for example a stereotactically placed catheter or a stereotactically directed laser beam – may be used to direct the surgeon from the surface, down an optimal trajectory to a deep seated tumour. This method, by reducing the operative trauma can increase the number of tumours which are accessible for removal.

Conclusion

Open stereotactic technique in the 1990's is essential for the treatment of a number of functional conditions and cystic space occupying lesions. It has an important part to play in the investigation of tumours and can help to increase the number which are accessible to treatment. It is regrettable that the technique is not, even today, in routine use in all neurosurgical departments. This is undoubtedly the result of a misunderstanding of the role of stereotactic technique for a modern neurosurgical service. It may also be the result of the innate conservatism of those who would have to finance the acquisition of the necessary equipment and of those who would have to use it, once acquired. In the opinion of the author, the increase in precision and the reduction of surgical trauma, inherent in the technique makes it very difficult to argue against its widespread, routine application. Nonetheless, it has its limitations. As yet it cannot of itself be used in the effective removal of solid tumours. However, it can be employed to guide not only solid instruments but also ionising irradiation to mass lesion targets. It is just this combination of stereotactic guidance and narrow beam, high energy radiation, to a precisely defined target, that is the basis of Gamma Knife radiosurgery.

Suggested Further Reading

1. Bosch DA (1986) Stereotactic techniques in clinical neurosurgery. Springer, Wien New York, pp 1–14

2. Gildenberg PL (1988) General concepts of stereotactic surgery. In: Lunsford LD (ed) Modern stereotactic neurosurgery. Martinus Nijhoff Publishing, Boston, pp 3–11
3. Horsley V, Clarke RH (1908) The structure and functions of the cerebellum examined by a new method. Brain 31: 45–124
4. Leksell L A (1949) Stereotaxic apparatus for intracerebral surgery. Acta Chir scand 99: 229–233
5. Spiegel EA, Wycis HT (1962) Stereoencephalotomy. Part II. Clinical and physiological applications. Gruve, New York
6. Spiegel EA, Wycis HT, Marks M, Lee AJ (1947) Stereotaxic apparatus for operations on human brain. Science 106: 349–350
7. Talairach J, David M, Tournoux P, et al (1957) Atlas d'anatomie stereotaxique. Masson, Paris

3. Ionising Radiation and Its Physical and Chemical Effects on Living Tissue

Introduction

Some knowledge of the effects of ionising radiation on living tissue is necessary, for those who wish to understand the nature of any treatment using radiation and who also wish to inform patients about such treatment. Correct information is particularly important in this regard, because of the associations that the word radiation has for most people. Most powerful therapeutic methods arouse concern or anxiety amongst those who will be subjected to them. However, medicine or the surgeon's knife may also be viewed with some degree of relief. This is not the case for treatment involving ionising radiation There are several reasons for this. Firstly, everyone has seen films of the agonising effects of radiation fall-out from nuclear bombs. Secondly, there is something particularly horrid about the idea of an invisible entity, creeping into the body, producing no apparent initial effect but followed by untold harm, at a later date. The unpleasantness is compounded by the nature of the harm that may develop: in that sterility and cancer are amongst the commonest consequences of exposure to excess irradiation. Thirdly, the concept of radiotherapy is indissolubly linked in the public mind with cancer. This last provides particular difficulties for the patient with an intracranial tumour, where radiation treatment may be advisable because of a tumour's inaccessibility, rather than its malignancy.

Radiophysics

Basic Concepts

The basic mechanisms, underlying the effect of radiation on matter occur at the atomic and subatomic level. Thus, understanding of these effects is necessarily predicated on some elementary knowledge of atomic theory. The concept of an atom, consisting of a nucleus containing a specific number of positively charged protons and non-charged neutrons, surrounded by orbits of electrons is familiar.

Another important set of concepts in the study of atoms and subatomic particles are embodied in a theory; Quantum Theory. Quantum theory was developed to explain findings which indicated that electromagnetic radiation (see below) sometimes appeared to behave as a wave and sometimes appeared to behave as a stream of particles or quanta, each carrying a certain specific amount of energy, defined by the Planck Radiation Formula:

$$E = hv$$

where **E** is energy, **h** is Planck's constant and **v** is the frequency of the radiation. Moreover, it is also true that subatomic particles may be considered to have wave like properties. Another important concept, for the understanding of radiation's interaction with matter, is that energy and matter are interconvertable, according to Einstein's famous equation:

$$E = mc^2$$

where **E** is energy, **m** is the mass of the particle being converted into or arising from energy and **c** is the velocity of light. This dual nature of radiation and subatomic particles is not intuitive and has been considered difficult to understand. However, it is not really so. Many familiar objects have different properties depending on how they are observed. For example, a meadow will be viewed by a farmer in terms of what he can grow on it. A geologist, looking at the vegetation and terrain may try to deduce if there is oil below. The archeologist will use other observations to deduce what artefacts may lie buried. It is the same meadow showing different properties depending on how it is observed. To return to quantum mechanics. Another analogy is to consider looking at a cylinder. When viewed from the side it can appear as a rectangle. When viewed from the top it can seem to be a circle. Obviously, it is both and neither. Moreover, in keeping with the Quantum conception of waves and particles, the two views of the cylinder are mutually exclusive. In the same way, a wave may seem to be a wave or a particle, but never both at the same time. Now we must return to quantum mechanics. Irrespective of whether waves or particles are considered, an atom can only emit or absorb energy in discrete, discontinuous quanta. The quanta so emitted or absorbed will have a particular energy and by the same token a particular wavelength; in accordance with Planck's equation.

The term **ionising radiation** refers to radiation which has a sufficiently high energy to be able to dislodge electrons from atoms, or disrupt the bonds between atoms and molecules. Examples of this sort of radiation are ultraviolet light, X-rays and gamma rays. An atom deprived of an electron will have a net positive charge and thus will have become an ion hence the term ionising radiation

There are two sorts of radiation **source** used in radiation treatment, artificially generated irradiation from man-made machines and spontaneously generated radiation from radio-nuclides. (A nuclide is a variety of an atom. It may be used as an alternative to the word isotope. It also applies for atoms which have no isotopes and exist only as a single variety; such as sodium or fluorine. A radionuclide is an unstable atomic variety exhibiting spontaneous radioactive breakdown).

There are two basic kinds of radiation in current use. *Electromagnetic radiation* has no mass and travels at the velocity of light $(c = 3 \times 10^8 \text{ m/s})$. *Particle radiation* consisting of for example protons, neutrons or electrons, has mass and travels at a lesser velocity. Both particles and electromagnetic radiation lose energy to matter by interacting with it. If a radiation passes through matter without striking an atom no ionisation will occur.

Units

A variety of units are used to quantify the phenomena associated with ionising radiation. The units relate to the expenditure of energy and common to a number of them is the SI unit for energy, the joule. A **joule** is the energy expended when a force of one newton works through one metre. A **newton** is a force which will accelerate a mass of 1 kg to 1 m/Sec^2. The following are among the more useful units used in relation to radiation therapy.

1. In the measurement of the energy of photons and particle beams electron volts are used.

The **volt** is the difference in potential between two points within an electric field which requires one joule of energy to move one coulomb of electric charge between the two points. A **coulomb** is the quantity of electricity transported by a one ampère current in one second.

1 **electron volt** (eV) is the energy acquired by an electron passing through a potential difference of 1 volt and is equal to 1.60219×10^{-19} Joules.

Kilo electron volts (KeV) and **mega electron volts** (MeV) are the customary units used in radiation therapy.

$$1 \text{ KeV} = 10^3 \text{ eV and } 1 \text{ MeV} = 10^6 \text{ eV}$$

2. An older out-dated set of units are **Kilovolts** (KV) and **Megavolts** are used. These are descriptive terms related to the quality and not the energy of the radiation. They are only mentioned because they occur commonly in older texts.

3. The **Activity** (A) of an amount of a radioactive nuclide is the number of spontaneous nuclear transitions (dV) in the time interval dt.

$$A = - (dV/dt)$$

The unit of activity is the **Becquerel** (Bq)

An older unit of activity is the **Curie** (Ci)

In the measurement of nuclear disintegration of a radionuclide:

$$1 \text{ Curie (Ci)} = 3.7 \times 10^{10} \text{ disintegrations per second.}$$
$$1 \text{ Bequerel} = 1 \text{ disintegration/sec.}$$

4. In the measurement of the tissue absorption of radiation: The absorbed dose, D, is the mean energy imparted, dE, by ionising radiation to matter, in a volume element with mass dm.

$$D = dE/dm$$

The unit of absorbed dose is the **gray** (Gy)

$$1 \text{ Gy} = 1 \text{ joule/kg}$$

1 Radiation Absorbed Dose (rad) is a familiar and older radiation absorption dose, using CGS units so that

$$1 \text{ rad} = 1 \text{erg/100g}$$

It should no longer be used.

The relation of rads to Gy is described by the equation

$$1 \text{ Gy} = 100 \text{ rads}$$

In the U.S.A. and the U.K. the cGy is popular.

$$100 \text{ cGy} = 1 \text{ Gy} = 100 \text{ rad.}$$

Electromagnetic Radiation

There is of course a vast range of electromagnetic radiations from the lowest frequency radio waves (frequency 10 kHz, wavelength 30 kilometres) up to cosmic rays (frequency 10^{24} kHz, wavelength

1 /1000 millionth of an Angstrom unit). In low frequency, long wavelength radiation the wave-like properties dominate. In high frequency, short wavelength radiation, the particle-like properties dominate. For the present purpose, the range of interest is X-rays (approx. 10^{15} to 10^{21} Hz) and gamma rays (approx. 10^{18} to 10^{24} Hz). The essential difference between X-rays and gamma rays is their manner of production. X-rays are the product of deceleration of a stream of electrons. According to electromagnetic theory, acceleration or deceleration of electrically charged particles results in the emission of radiation. This phenomenon is used in the production of X-rays from a machine, in which negatively charged electrons are accelerated in a vacuum and strike a target with a high atomic weight, usually tungsten. The X-rays produced by the deceleration of the electrons may diffuse in many directions and are directed where desired through a device called a **collimator.** The X-rays produced by different machines can have different energies. **Linear Accelerators,** much used in conventional radiotherapy, and which may be adapted to perform a sort of radiosurgery, produce X-ray beams with energies between 4 and 30 MeV.

It is important not to confuse the **energy** of a radiation beam with its *intensity* (see the Planck Radiation Formula). The energy of the beam is proportional to the frequency of the radiation. It will be seen later that this is in turn related to the penetration of the beam and the sort of effects it has on the atoms which it strikes. The intensity of the beam is related to the *number* of photons the machine is delivering; in other words to the dose. This may become clearer by considering the same characteristics in respect of a beam of light. Increasing the energy, or frequency of a beam of light will change its colour from red through the spectrum up to violet. For a particular colour, increasing the intensity will make the light brighter. It will not change the colour.

Gamma rays, which are also electromagnetic radiations differ from X-rays in that they are produced in a different way (Fig. 3.1). When the nucleus of an atom is in an excited state it can decay to a stable state by emission of one or more photons, (quanta of electromagnetic radiation energy) called in this case gamma rays or gamma photons. The other products of radioactive breakdown are alpha particles (helium nuclei with 2 protons and 2 neutrons) and beta particles (electrons) and neutrons. Alpha particles produced by radioactive transformation have too low a penetration to be of much use in clinical practice. Beta radiation also has a low penetration but it may be used following implantation of isotopes in tissue. One example, in the field of neurosurgery is the highly successful

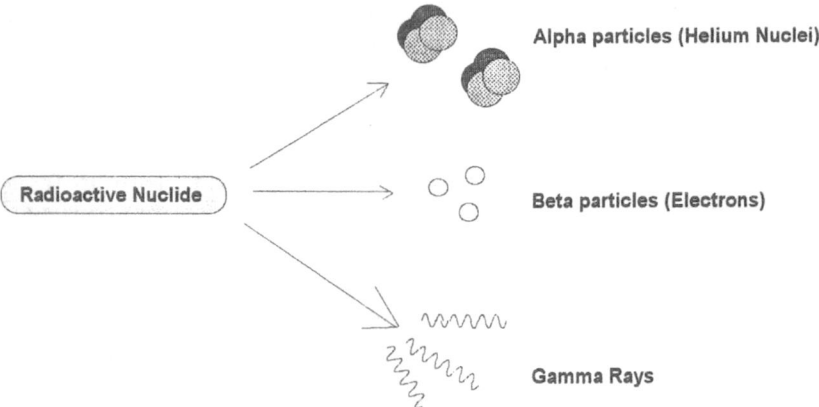

Fig. 3.1 Radioactive Decay
There are three main types of radioactive breakdown, producing three main products as illustrated here. However, not all radioactive breakdowns produce all products. Thus alpha particles are mostly produced by isotypes of high atomic weight, for example Uranium. Cobalt-60 produces gamma waves, but also loses electrons

treatment of craniopharyngioma cysts with instillation of radioactive Yttrium, which emits pure beta radiation. This is not unusual. Many nuclides emit a greater proportion of one of the products of radioactive breakdown than of the others. Thus Uranium-235 is primarily an alpha emitter, just as Cobalt-60 is mainly a gamma emitter but also emits beta particles.

Particle Radiation

The most commonly used particles in current therapeutic use are electrons and protons. Electron beams can be produced by an adapted linear accelerator or a special sort of accelerator called a betatron. High energy electrons have a definite limited range in tissues, with a rapid fall of dose over distance. This makes this form of radiation advantageous in the treatment of cutaneous and subcutaneous lesions, such as lymph nodes or the parotid gland. The commonest energies of electron beams are 7 to 18 MeV, for the linear accelerator and 12.4 to 124 MeV for a betatron.

Proton particles are produced in particle accelerators, such as the synchrocyclotron. These are rare and costly and are found in only a few centres, for example Harvard, Boston, USA and Uppsala in Sweden. Particle beams have special characteristics, enabling the delivery of a sharply defined dose deep in the tissues, with relative sparing of the tissues on the way in to the high dose volume. The energies of those in current use ranges from 72 to 1000 MeV.

The Effect of Electromagnetic Radiation on Matter

Ionisation is a chemical change. Electromagnetic radiations can react with matter in a variety of ways; for example reflection, refraction interference, that is different forms of scattering. They can induce chemical change only by *absorption*. When ionising radiation is absorbed it interacts with atoms to detach electrons from their orbits. The energy of these electrons is part of the energy of the incoming photons. There are three main ways in which such interactions between radiation and matter occur, depending on the energy of the radiation. Finally, it should be repeated that some radiation will go through whatever matter is being irradiated without interacting with it.

The Photoelectric Effect

This is the major energy absorption mechanism for low energy X-ray beams up to 50 keV, though it also occurs at higher energy levels. All the energy of a given photon is absorbed in detaching an electron from one of the inmost shells of an atom. An outer electron will then hop into the insufficiently filled inner shell, resulting in a change in energy level and the emission of a photon of X-rays (Fig. 3.2). The kinetic energy of the originally ejected inner electron will be equal to the energy of the incident photon minus the energy required to detach it from its orbit. It is *not* related to the *intensity* of the photon beam which will determine the *number* of detached electrons. These liberated electrons will then cause the ejection of further electrons from other atoms, until their kinetic energy is no longer great enough to change an atomic energy state. At this stage they can produce vibrations within atoms but not ionisation.

Compton Scattering

With higher energy X-ray beams and gamma rays, with an energy between approximately 90 keV and 5 MeV, a different effect occurs; involving the interaction of the radiation photon with electrons in the outer shell of the atom. Some of the photon's energy will be dissipated in detaching the electron from its path and in giving it kinetic energy. The rest of the energy will continue as a new photon with an energy

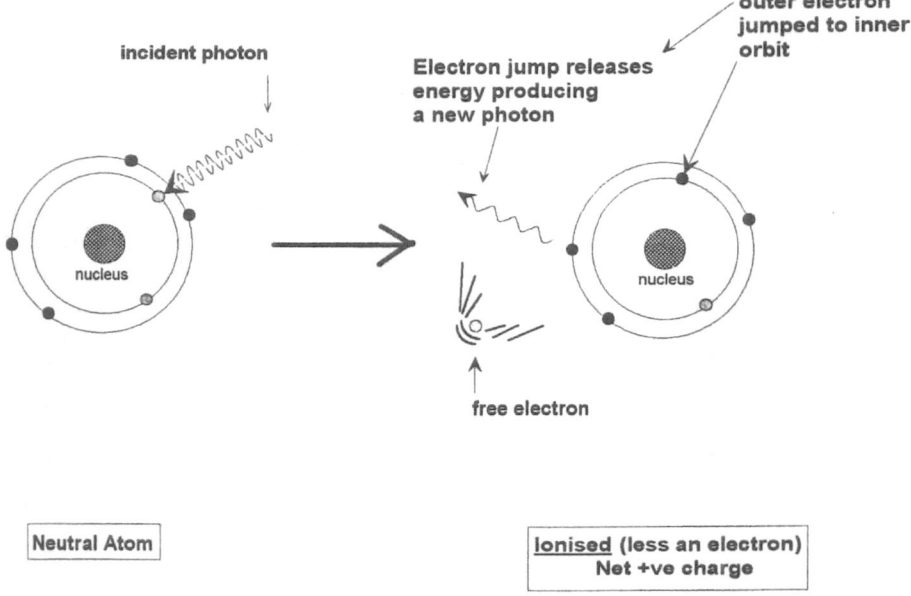

Fig. 3.2 Photoelectric Effect

The photoelectric effect was discovered following the observation that when a spark appears at a gap between two electrodes its appearance could be facilitated by shining light at the gap. The spark was due to electrons crossing the gap. The photoelectric effect is due to the incoming photon from the light loosening an electron and thus facilitating the generation of a spark. The electron released is from the inner shell. The energy of the ongoing photon is discharged when an electron jumps from an outer to the inner electron shell. The energy of this photon is equal to the energy of the incoming photon less the energy imparted to the free electron. Note that the frequency of the incoming photon is less than that of the ongoing photon. This reflects the relationship between the energy and frequency of a photon. See text

equal to the energy of the incident photon less the energy required to detach the electron and the kinetic energy delivered to that electron (Fig. 3.3). This new photon with a lower energy will naturally have a longer wave length. Quantitative analysis of the effects of radiation in water has shown that the vast majority of the energy absorbed is related to the detached electrons and not to the ongoing lower energy photons.

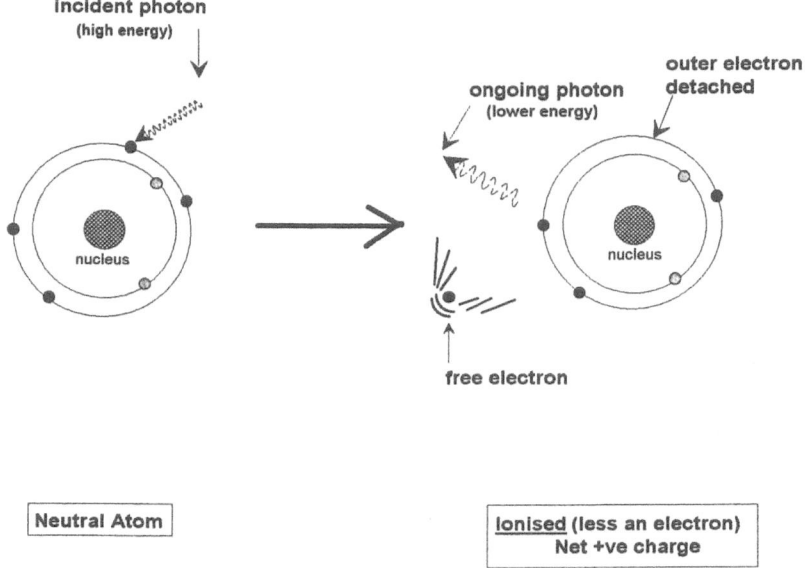

Fig. 3.3 Compton Scattering

Compton found that when X-rays are dispersed in a crystal, there was a change in the frequency of the X-rays, indicating a loss of energy. At the same time an outer electron is freed. The lower frequency of the ongoing X-rays is equal to the energy of the incoming X-ray photons less the energy imparted to the ongoing electron. Note that, as in the photoelectric effect, there is a net gain of one positive charge, and the atom is ionised. This time the energy liberated comes directly from the incoming photon. It is not mediated by means of an electron hopping from one orbit to another. It is the most likely process to be responsible for ionisation during Gamma Knife surgery

Pair Production

When a photon passes close to the nucleus of an atom it is exposed to the powerful energy field around that nucleus and may thus be converted from a photon of energy into matter, in the form of a pair of electrons. Since the mass of an electron is equivalent to 0.511 MeV the energy of the incident photon must be at least 0.511×2 or 1.022 MeV. One of the pair of electrons has a positive charge (positron) and the other a negative charge (electron). Both these electrons pass through the absorbing matter exciting and ionising atoms as described earlier (see Fig. 3.4).

The different mechanisms of energy absorption are not mutually exclusive: though the coexistence of the photoelectric effect and pair production is not thought to occur. However, the radiation energy

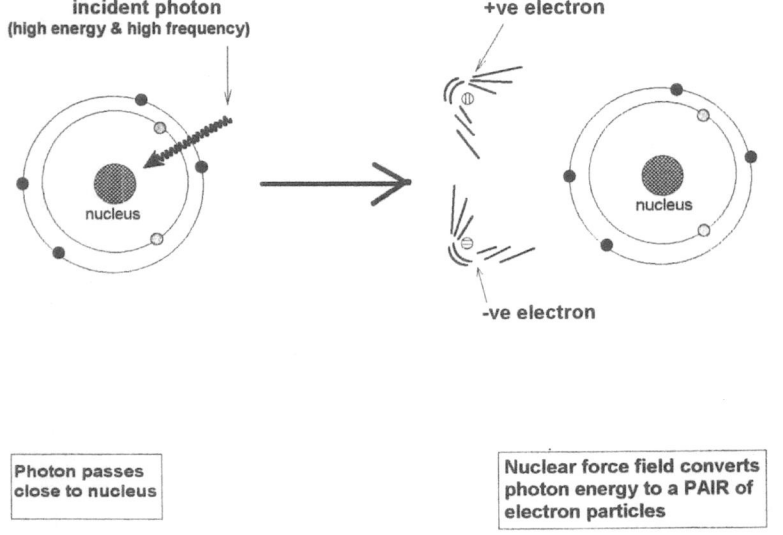

Photon passes
close to nucleus

Nuclear force field converts
photon energy to a PAIR of
electron particles

Fig. 3.4 Pair Production
In this case, the incoming photon has a sufficiently high energy to reach close to the nucleus. Here it is affected by the powerful field and is converted into energy. Two electrons are formed in accordance with the law of Conservation of Electric charge. Thus the minimum number of electrons that can be formed in this way is one with a positive and one with a negative charge. In this way electric neutrality is maintained, and the law is obeyed

range, associated with Compton scattering will at its lower end also be associated with the photoelectric effect, while at its upper range it will be associated with pair production.

The Effect of Charged Particles on Matter

A crucial difference in the pattern of energy absorption between particles and electromagnetic radiation is related to this characteristic of particles: that they can decelerate while radiations are bound to travel at the speed of light.

The absorption of energy associated with the passage of particles through matter is described by the Linear Energy Transfer (LET), described by the formula - dE/dX where dE is energy loss and dX is unit distance travelled.

The units of LET are KeV/mu: where 1mu = 10^{-6}m.

The energy loss of a particle is reflected by ionisations along the course of its passage. How far the particles will travel in a medium – such as living tissue – is a function of the density of the medium and its atomic weight on the one hand, and the mass and velocity of the particle on the other. Protons as an example of "charged heavy particles", with their greater mass can penetrate more deeply. There is relatively little energy loss along the track of a proton beam, so long as the particle is moving quickly. Thus, such heavy particles have a low LET in the part of their tracks where they are moving fast. However, more and more of their energy is absorbed as they decelerate, so that this part of the track has a high LET. Since most energy absorption occurs at the distal end of the track, most of the ionisations also occur in this region. A consequence of this phenomenon is that particles like protons, with an appropriate delivery system can be used to produce very precisely defined radiation fields at specific distances from the particle source. The precisely defined area of intense irradiation at the end of a low LET track following the passage of protons, is called a "Bragg Peak". However, taking advantage of the Bragg Peak phenomenon is not the only way in which a proton beam may be used to produce a well localized volume of high radiation energy delivery. **Cross firing** of a number of *narrow* proton beams will also produce a region of high dose where the beams cross, while the amount of dose delivered along the beam outside the cross firing region will be low because of proton radiation's low LET. To avoid the development of a Bragg Peak a proton source is used with a high energy and therefore a high penetration, so that the deceleration of protons, necessary for a Bragg Peak will occur after the protons have passed through the living tissue and emerged on the far side (Fig. 3.5). This principle for producing focused ionisation is mentioned again in the chapters dealing with the development of the Gamma Knife. Obtaining a sharply defined radiation dose, by using cross firing of a number of narrow radiation beams is a central principle of Gamma Knife radiosurgery.

The principles of particle radiation applying to protons apply also to electrons, but because of their lower mass they penetrate less well into tissues, at comparable energies and are most useful for the discrete irradiation of superficial structures.

In conclusion, it may be noted that irrespective of the kind or dose of radiation energy delivered to a substance, the majority of the energy absorbed by the cell is mediated through the free electrons, produced by the radiation.

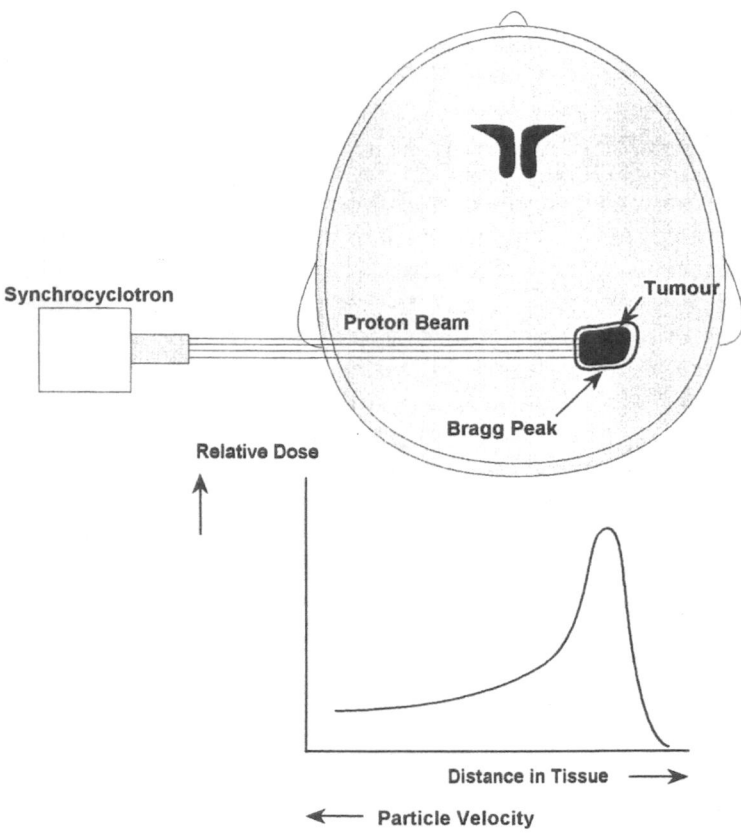

Fig. 3.5 Focal Radiation (Bragg Peak)
Most radiation energy delivered from particle radiation is lost when the particles decelerate. This deceleration of particles enables the concentration of the radiation dose over a tightly controlled sharply defined volume, called the Bragg Peak. This is one of the best known techniques for delivering focused radiation today. For the sake of presenting the picture of the radiation and the graph of particle deceleration concordantly, the source in this diagram is placed on the opposite side from the lesion. This would not really happen in the clinic

Radiochemistry

Early Effects

The events described in the previous section relate to the effects of radiation energy on atoms – the so-called **direct effect** of radiation. These events take place within *fractions of a microsecond*. As stated above, a common feature of all forms of radiation is the production of free electrons moving at speed through the absorbent medium. These can combine with ions of the same sort as those from which they were derived. They can also combine with atoms of other molecules, producing energised unstable products. This is the basis of the **indirect effect** of radiation. In living tissues, among the available molecules which can thus react with electrons is the water molecule, which is present in abundance. In a *matter of microseconds* free radicals can form according to, for example, the following equations.

$$H_2O \leftrightarrow H_2O^+ + e_{aq}^-$$

$$H_2O^+ \leftrightarrow H^+ + OH$$

$$H_2O + e^- \leftrightarrow H^+ + OH^-$$

Other possible products of this kind of reaction are H, H_2, H_2O_2. Which radicals are formed and in which direction the equations are predominantly balanced will depend on the nature of the absorbant medium, including its physical state and also the energy of the radiation. These free radicals can then react with other molecules, within the cell to do damage. These chemical reactions are enhanced by the presence of oxygen. Or perhaps it is more precise to say, that if hypoxia induces a tissue pO_2 of less than 30 mm Hg, then these reactions are inhibited and the damaging effect of the radiation is reduced. Because the half life of the free radicals is so short, they cannot travel great distances and the damage they induce is effectively limited close to the path of the radiation and does not spread to any great extent through the tissues.

Everything that has been described so far in this chapter would be of little interest, if the effects of radiation were consistently reversible. However, they are not, because the energy that is deposited within living cells proceeds, also within *microseconds* to damage the macromolecules which are the substrate of living processes. Chemical bonds may be broken, polymers may depolymerize or new unphysiological polymerization may occur. These processes may be reversible and be subsequently repaired, but they may be permanent leading to **biochemical injury.** The biochemical changes occur over the *seconds to hours* following irradiation and will be considered in the next chapter. It is generally thought that the most important target for biochemical injury is the hereditary molecule DNA.

Radiation Injuries to Nucleic Acid Molecules

The question then arises as to why DNA is considered to be the most likely target for radiation injury. There are three main lines of evidence. Firstly, selective irradiation of nuclei and cytoplasm separately has shown that the nucleus is far more susceptible. Secondly, cell death is more easily achieved, when binding radioactive nuclides to different intracellular macromolecules, if the said nuclides are incorporated into the DNA. Finally, there is a strong correlation between the radiation responsiveness of a cell and its DNA content.

There are four major sort of DNA injuries which are considered to be the most common caused by radiation. These are base damage, cross links with nuclear proteins, single DNA strand breaks – **SSB** - or double DNA strand breaks - **DSB** (Fig. 3.6). The evidence suggests that it is those DSB, which do not repair after several hours, which are responsible for the sterilization of the cell. It would seem that 1 Gy of radiation produces about 2000 initial single strand breaks and about 40 double strand breaks. However, the lesion that counts is the DSB which is not repaired.

Repair of Radiation Damage to DNA Molecules

It has also been reckoned that 1 Gy produces about one lethal injury per cell, so that most of the damage inflicted may be repaired; indicating the effectiveness of the repair processes. Oxygen can affect the reparation process both by binding to DNA chain breaks and making them permanent and by reacting to accelerate the processes of repair. Thus, the role of oxygen is complex and it bears repeating that the inhibition of radiation damage by hypoxia has been shown to apply, only if the tissue pO_2 is less than 30 mm Hg.

Biological Changes

So far this chapter has consisted of an outline of how ionising radiation can damage living tissue, physically and chemically, at the subatomic and molecular level. These chemical and physical changes will be expressed biologically in two main ways. The membranes, enzymes and protein factory of the affected cells may cease to function or function in a deranged fashion. The reproductive functions of the cell may be damaged with destruction or damage to chromosomes, delay in mitosis, mutation and changes in the cell cycle. Reflecting the possibility for repair at the physical and chemical levels, these biological changes may recover or they may not. Finally, if the above changes are not lethal in the short term, over a time scale of months to years late effects may be seen in the form of premature aging, carcinogenesis or growth disturbances in the young.

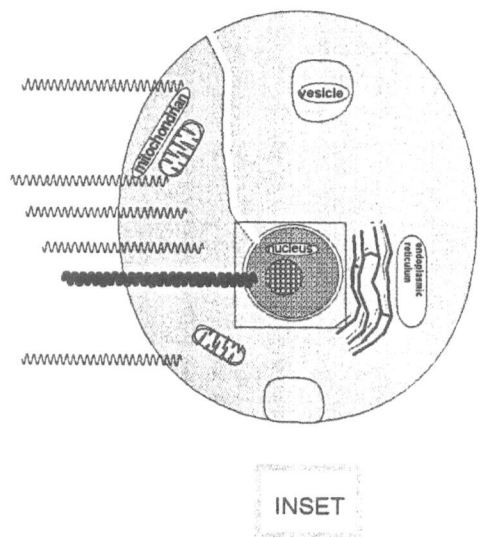

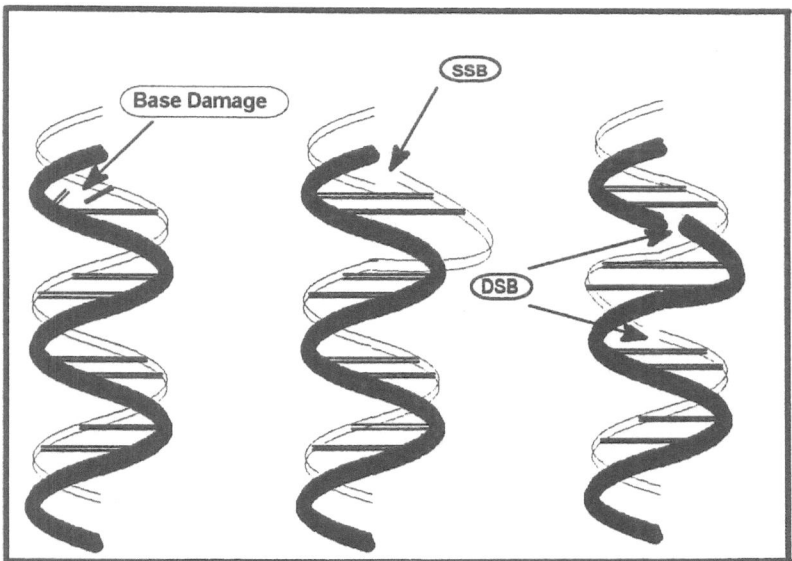

Fig. 3.6 Cellular Radiation Targets

Radiation can strike anywhere in a cell, but nuclear DNA is seen as the most important target, as indicated here by the thicker radiation penetrating the nucleus. The inset shows 3 of the common forms of DNA damage. About 2000 single strand breaks (*SSBs*) occur for 1 Gy, which at the same time produces about 40 double strand breaks (*DSBs*). It is DSBs which are the lethal lesion. But 1 Gy is also considered to produce an average of 1 lethal lesion per cell, indicating the efficiency of the repair processes

Suggested Further Reading

1. Larsson B, Lidén K, Sarby B (1974) Irradiation of small structures through the intact skull. Acta Radiol 13: 512–534
2. Perez CA (1977) Principles of radiation therapy. In: Horton J, Hill GJ (eds) Clinical oncology. WB Saunders Co., Philadelphia, pp 126–141
3. Sampiere VA (1980) Radiation measures and dosimetric practices. In: Fletcher GH (ed) Textbook of radiotherapy. Lea & Febiger, Philadelphia, pp 1–40
4. Withers HR, Peters LJ (1980) Biological aspects of radiation therapy. In: Fletcher GH (ed) Textbook of radiotherapy. Lea & Febiger, Philadelphia, pp 103–180

4. Biological Effects of Ionising Radiation

Introduction

The effects of radiation at atomic and molecular levels have been discussed in the previous chapter: events that occur over a time frame of nanoseconds to milliseconds. The present chapter relates to the effects of radiation on visible structures, in other words cells and tissues: events that occur within a time range of seconds to months or even years. To begin with, the theories relating to the infliction of radiation damage at the cellular level must be mentioned. It is considered that this may occur in one of two main ways. There is the *Direct Action* theory, whereby the ionisation of and lesion to the target, most probably DNA, is the primary event. The *Indirect Action* theory refers to the formation of DNA lesions produced by free radicals, which have in this instance been the primary target for ionisation by the radiation. Experimental work shows that the effect of oxygen on radiosensitivity is mediated by indirect action, since the tissue pO_2 affects the formation of the free radicals, which damage the DNA. Today, both modes of action are considered to be important in producing cell death. It may be mentioned that while there is, as stated above, broad agreement that the most important *site* of damage is the DNA of the cell nucleus, it may not be the only one. Some radiation effects, for example radiation oedema, indicate that cell membranes may also be important targets.

Cell Survival Studies

Scientific analysis of any phenomenon requires quantitation, as an aid to improving understanding of that phenomenon and radiobiology is no exception. This involves the selection of variables to be observed and the registration of end-points. One of the most useful of radiobiological variables to study has been cell survival.

Cell survival studies are for the most part based on the effects of radiation on cells in culture, in other words *in vitro*. It must also be emphasised that cell death in the present context has a particular precise definition. **Cell death is the loss of the capacity for <u>indefinite</u> proliferation.** The methods for studying cell survival, following a dose of radiation are designed with this definition in mind. Various models are used, from which the cells are plated out and the colonies of dividing – thereby surviving – cells are counted. Such surviving cells are termed *Clonogenic Cells*. The proportion of surviving cells is called the *Survival Fraction* and is a much used quantitative indicator of the effect of the radiation.

To appreciate the significance of these survival studies, it must be understood that the chance of radiation producing a cell kill is a random event. Thus, if a group of cells receive a dose which is just big enough to provide one lethal event per cell, not all cells will be killed. Some will receive no lethal dose and some will receive several. The type of statistics used in calculating the chance of a cell kill are called Poisson statistics. This statistical technique is employed when the chance of a specific given change being produced is very small, in relation to the number of events taking place. This method is obviously relevant in the present context, since it is estimated that 1 Gy of radiation will give rise to 10^5 ionisations per cell. On the other hand this amount of radiation produces only about 40 double strand breaks in the cell's DNA. Using Poisson statistics it is calculated that the percentage of cells which survive, when a radiation dose, sufficient to produce *an average* one lethal lesion per cell is delivered, is e^{-1} or 37%. This is called the *Survival Fraction*. The dose producing a survival fraction of e^{-1} is called D_0. The relationship between the survival fraction and the dose can be expressed as an equation. The form of such an equation should reflect the underlying pathophysiological mechanisms involved in cell death. Over the years a number of equations have been derived which have all been more or less complex exponential functions. They have been derived on the basis of the concept of *Target Theory*. This theory proposes that there exists a specific number of targets in the DNA and these must be damaged if cell death is to occur. The above-mentioned exponential equations have all been shown to fit the actual observed patterns of cell survival inadequately. Thus, in the last 10 years a different mathematical formalism, the linear quadratic equation, has been increasingly used.

The Linear Quadratic Equation

The linear quadratic equation has the form

$$p = e^{-(\alpha D + \beta D^2)}$$

where p is the survival fraction, α und β are constants, descriptive of the linear and quadratic components of the equation respectively (Fig. 4.1).

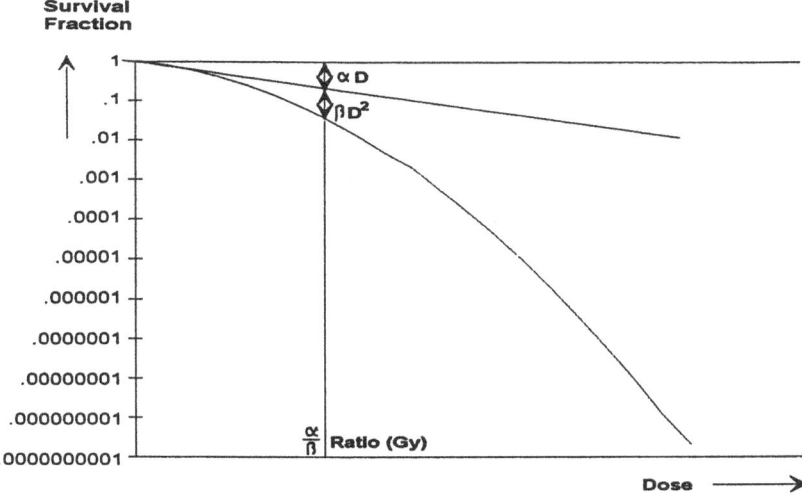

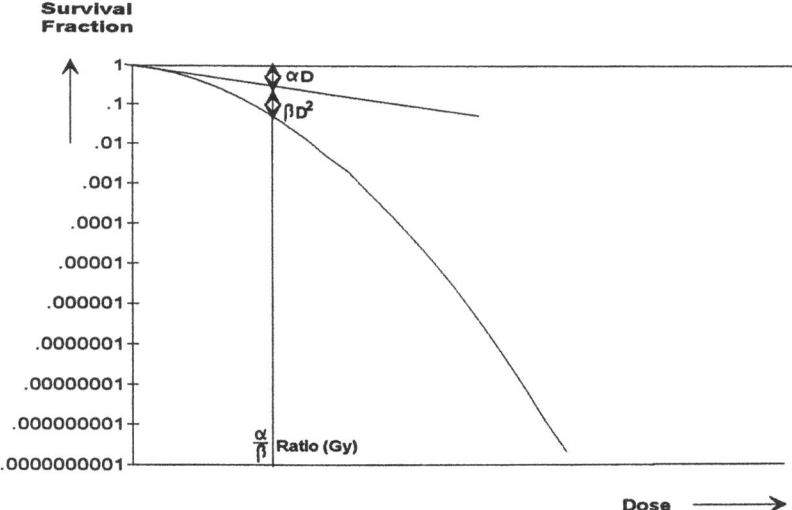

Fig. 4.1 Linear Quadratic Survival Fraction/Dose
A linear quadratic curve best describes the relationship of cell death to radiation dose. The cell death for a specific set of circumstances can be characterised by the α/β ratio, the dose in Gy at which the linear and quadratic parts of the curve contribute equally to the death of cells. A low α/β ratio indicates a more radioresponsive tissue. Other correlations are described in the text

The applicability of the linear quadratic formula carries with it certain implications.

Firstly, it is necessary to consider possible pathophysiological mechanisms which are both consistent with the formula and at the same time are consistent with the observed data of cell killing by radiation. The formula is presumed to reflect cell death produced by lesions, which are not related to any notional number of pre-determined targets. The shape of any cell survival curve will be determined by the α/β ratio. This defines the dose at which the linear and quadratic contributions to cell killing are equal. The units of the α/β ratio are Gy. The linear quadratic equation has over the last ten years not only been used for assessment of in vitro studies but has been widely applied in vivo, in the clinic.

There are two main conceptual models which, it is postulated may explain the form of the equation. The first is called the *Lethal-Potentially-Lethal* model (Fig. 4.2). In this model, attention is focused

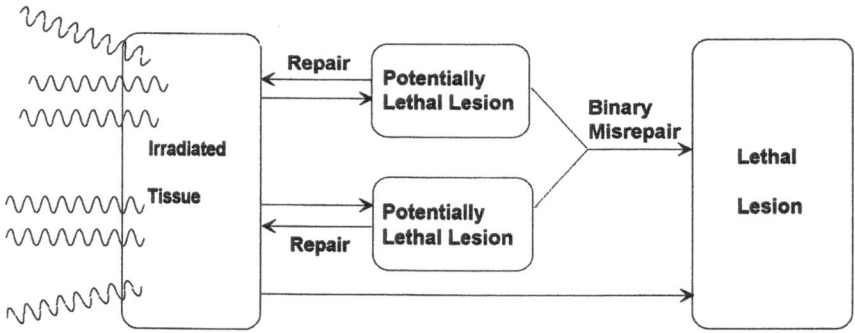

Fig. 4.2 Lethal – Potentially Lethal Model of Cell Killing
A biological model, consistent with a linear quadratic relationship between radiation dose and cell death, is necessary if the said relationship can have any value. One such model is the Lethal – Potential Lethal model. This concentrates on the lesion production and the interaction of lesions, in themselves insufficient to be lethal. The immediately lethal part of the model could relate to the linear part of the curve while the interaction of potentially lethal lesions could relate to the quadratic part of the curve

on patterns of lesion production. The linear part of the equation reflects lesions which are in themselves lethal. The quadratic part of the equation reflects lesions which of themselves may be reparable, or potentially lethal. While such lesions may repair, they can combine to produce lethal lesions. Thus, with increasing the dose, it is conceived that a lethal outcome of such lesions will increase according to a quadratic function. The second model is called the *Saturation Repair* model (Fig. 4.3). It suggests that the main biological substrate of the

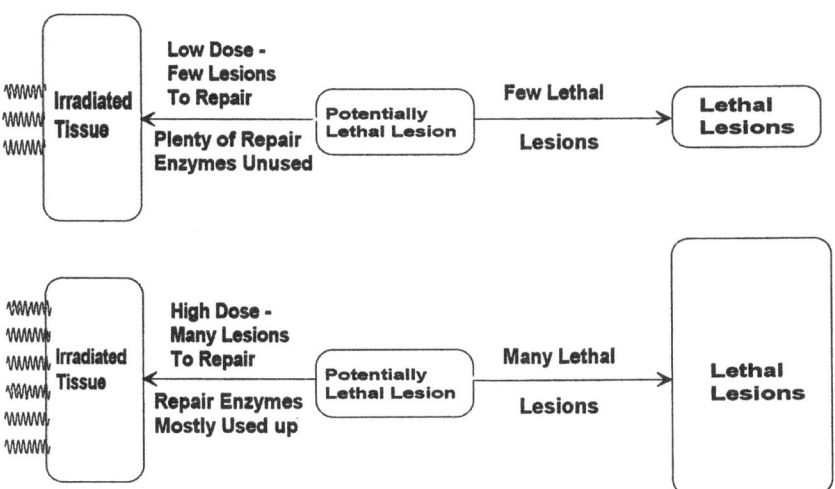

Fig. 4.3 Repair Saturation Model of Cell Killing
In this diagram another model is illustrated. It is also consistant with a linear quadratic relationship between radiation dose and cell death. This concentrates entirely on the accumulating effect of increasing radiation dose on the enzymes responsible for repairing the damage done by radiation. In the upper part of this diagram the dose is low while in the lower part it is higher, indicating a gradual failure of repair capacity. Thus direct hits producing immediate lesions could account for the linear part of curve. The quadratic part of the curve could reflect the increasing failure of overloaded repair enzymes with increasing radiation dose

linear quadratic equation relates exclusively to the processes of repair. Thus, with increasing doses an increasing number of potentially lethal lesions will be produced. With a greater number of lesions to repair, the repair enzymes and processes may be saturated by the number of repairs to be performed. Thus, a greater number of potentially lethal lesions end up lethal. It is conceivable that both models may contribute to the outcome, following radiation.

It is necessary to specify that *repair* and *recovery* do not mean the same thing when applied to the processes within a tissue, following a dose of ionising radiation. *Repair* refers to the intracellular restitution of damaged structures, in particular DNA. *Recovery* refers to the way in which the repair processes express themselves in the tissue; for example in terms of increased cell survival or reduction in the extent of radiation damage.

There are other factors to consider, in addition to the basic interaction between radiation and cells that has been discussed so far. These factors relate both to the tissue and to the radiation.

Different Factors Affecting the Effects of Radiation

Biological Factors

The Cell Cycle

Under the microscope, the only time that the DNA is visible is during mitosis. However, using autoradiography techniques, the DNA can be shown to pass through a cycle of changes. The M phase is mitosis. This is followed by an inactive phase or gap, the G_1 phase which is turn is followed by the S phase or phase of active synthesis. The S phase is in turn followed by a second gap, the G_2 phase which ends with the next

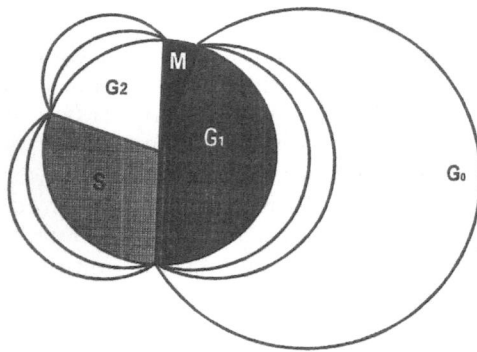

Fig. 4.4 Cell Division Cycle
The genetic material in a cell passes through a sequence of phases. *M* is the phase of mitosis. G_1 is the first resting phase. *S* is the phase of synthesis of new genetic material. G_2 is the second resting phase. *Go* is a G_1 phase that is so prolonged as to represent an almost total stop in the changes to genetic material. The cycle only applies to such material. The rest of the cellular activities continue as normal. A variety of circles are depicted for the various phases to indicate the variety that can exist between different cell types. The G_1 and *S* phases are relatively radioresistant while the G_2 and *M* phases are relatively radiosensitive

mitosis. Figure 4.3 depicts a variety of cell cycles. A variety is shown to indicate the variability between cells. There is also a Go phase which is a state of apparent suspended animation, when the cell cycle is either resting or changing so slowly as to be undetectable. During this phase, the rest of the cell's activities, that are not related to cell division continue unabated. Since normal adult tissues do not grow, the process of synthesis described by the cell cycle must also be accompanied by a process of cell loss, to provide the steady state of unchanged tissue size. In tumours, this steady state is disrupted. Figure 4.4 depicts the effects of position in the cell cycle on radiosensitivity, using cell survival study curves.

Oxygenation

This has already been mentioned a number of times. Hypoxia, at levels of pO_2 below 30 mm Hg reduces the development of damaging free radicals and thus the degree of radiation damage (Fig. 4.5). Moreover, there is experimental evidence which suggests that hypoxic areas of a tumour may reoxygenate, during fractionated radiotherapy, due to a variety of factors. These include reduction in oxygen consumption by dead cells, reduction in the number of cells in relation to area of the capillary bed and reduction of the intratumoural pressure, permitting reopening of the microcirculation. Since normal tissues are well oxygenated in relation to tumour tissue, reoxygenation of tumour tissue theoretically improves the therapeutic index (see Fig. 5.1).

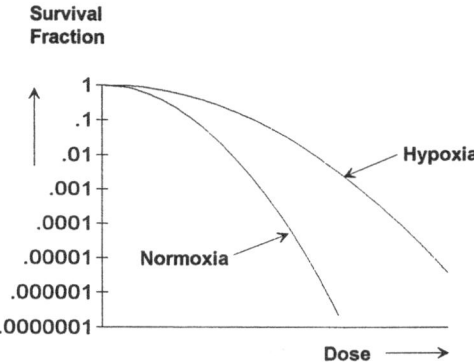

Fig. 4.5 Effect of Hypoxia on Dose Response
Radioresponsiveness is dependent on adequate oxygenation. If the tissue oxygen tension falls below 30 mm Hg the radiation is noticeably less effective, as shown in this diagram

Radiation Factors

Fractionation and Fraction Size

In conventional radiotherapy, it is customary to deliver the dose in fractions. This permits a higher target dose while keeping the normal tissue damage down to acceptable levels. It was originally based on the presumption that normal cells repair sublethal damage more quickly than tumour cells. It also has some other theoretical advantages In principle, it increases the chance of catching a tumour cell at a radiosensitive stage of the cell cycle (see above), by killing sensitive cells early and establishing cell cycle synchronisation; thus facilitating synchronisation of the cells at a sensitive phase, later in the treatment. This process is called reassortment. However, it may be seen that increasing the number of fractions increases the tumour cell survival, or decreases the proportion of cells killed. Figure 4.6 depicts the effect of fractionation on the total dose required to produce the same degree of cell death as shown using survival fraction studies. Thus, it is a general principle that to produce the same dose effect, using fractionation, a greater total dose is required than would be needed if all the radiation were given at one time. This is because of the capacity for repair of DNA and repopulation by undamaged cells, that occurs between the individual fractions, which make up the total dose. The

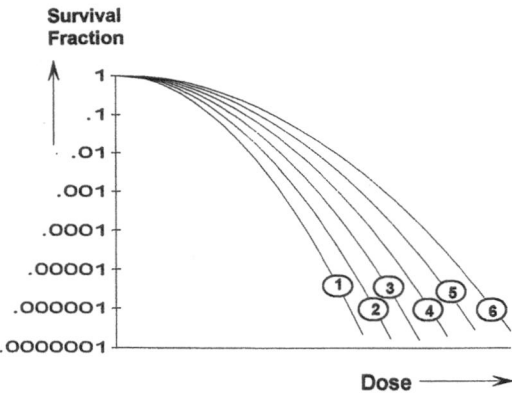

Fig. 4.6 Fractionation Effect on Dose Response
The biological effect of a given dose of radiation is less if it divided between several fractions. A corollary of this is that with an increasing number of fractions an equivalent increasing dose will be required to achieve the same amount of cell killing. This is illustrated diagrammatically here

phenomena described in this paragraph have been summarized as the 4 'Rs' of radiotherapy, **R**epair, **R**eassortment, **R**epopulation, and **R**eoxygenation. To these a fifth **'R'** may be added, **R**adiosensitivity, indicating the intrinsic sensitivity of individual tissues to radiation. This term must be distinguished from *radioresponsiveness*, which does not refer to a fundamental biological characteristic but to the clinical responsive to a given radiation treatment. It may be noted that of the 4 'Rs', repair and repopulation should reduce the effectiveness of the radiation with fractionation, while reoxygenation and reassortment should increase it. Finally, the linear quadratic equation has proven to be useful in determining appropriate fractionation strategies for different clinical situations and a little more on this will be mentioned in the next chapter.

Time Between Fractions

This is another radiation therapeutic variable that must be taken into account when fractionated doses are given. It must be obvious that two fractions given with one hour's interval and two fractions given with a month's interval are not equivalent. This becomes even more apparent when it is remembered that single band DNA damage repair is for the most part completed in 2 to 4 hours. Thus, in general, the first 5 to 6 hours are important: thereafter the time interval is not so significant.

Dose

Dose, as stated earlier, equals energy absorbed per unit mass of absorber, (dE/dM). Yet the term dose is, if unqualified, unacceptably imprecise. According to Webster's Collegiate Dictionary, the noun *"dose"* derives from Middle English, which derives from Old French which in turn derives from a Greek noun *"dosis"*, a giving: this noun being a derivative of the verb *"didonai"*, to give. However, there are a variety of different types of dose definition as outlined below, lending greater precision to our understanding.

The **Surface Dose** is the dose absorbed in the superficial layers of the skin.

The **Maximum Dose** is the maximum dose absorbed.

The **Target Dose** is, in general, the minimum dose delivered to the target volume.

The **Integral Dose** is the total absorbed dose in a specified volume (of tissue).

Isodose Curves are two-dimensional representations of the distribution of radiation, equivalent to isobars on a weather map or height contour lines on a survey map. The isodose lines pass through points of equal relative dose. They express relative dose distribution as a percentage of the maximum dose. Moreover, they reflect dose homogeneity.

The dose to the tumour is designed to be adequate for its destruction without producing unacceptable damage to the surrounding normal tissues. For fractionated doses to a brain tumour, a target dose of about 50 Gy has been found to be acceptable.

Dose Homogeneity

In conventional radiotherapy, radiant energy is delivered in one of two ways. With **teletherapy,** radiation is beamed into the patient from a distance. This technique enables the construction of radiation fields with a high degree of homogeneity over the target. This should be highly desirable, because the radiation effect cannot be greater than that due to the lowest dose delivered. **Brachytherapy** is the other principle radiotherapy technique and here a radiation source is implanted within the target. Lack of homogeneity is an intrinsic weakness of brachytherapy for solid tumours, though the degree of inhomogeneity can be reduced by using multiple radiation sources. Nonetheless, intracystic instillation of an isotope with limited penetration provides an excellent way of delivering an adequate dose to the walls of a thin walled neoplastic cyst. The cyst fluid diffuses the isotope, evens out the dose and is radioinsensitive, so that the dose at the wall is the only factor which needs consideration.

Dose Rate

Dose rate is the dose per unit time. Within certain limits dose rate appears to affect the biological effectiveness of a given dose. Between 1 Gy per minute and 1 Gy per hour there is a gradual fall off in dose effectiveness. Increasing or decreasing the dose rate beyond these limits has little extra effect. Dose rate calculations become important when the radiation source is an isotope, where the age of the isotope has reduced the dose rate below 1 Gy per minute. The effect of dose rate on radiosensitivity is yet another variable, which may be illustrated using cell survival studies.

Dose Volume

It has long been appreciated that the volume of a tumour is an important determinant of the success of radiotherapy on that tumour. In general, larger tumours do less well than smaller. On the other hand, it is known that for the same dose, the larger the area of skin receiving the dose the greater the chance of skin complications from that dose. Dose volume may be defined, for convenience, as the volume within a given isodose. Its significance may be demonstrated as follows. While 5 Gy to the whole human body is usually lethal, 5 Gy to a tumour volume is wholly inadequate. Indeed doses of up to 50 Gy can be tolerated with gastrointestinal cancer, while it is the gastrointestinal tract which is one of the main victims of whole body radiation. Moreover, considerations of volume must be made within the bounds of a single species. Some insects and bacteria can survive radiation doses vastly in excess of those tolerated by mammals.

Clinical Radiobiological Correlates

The Fate of the Normal Tissue

The fate of normal tissue, subjected to ionising radiation is closely related to the cellular architecture and internal organization of the tissue. In general, tissues are considered to have two main kinds of organization.

Hierarchical Tissues (H-type)

In these tissues, the cell division is carried out by a minority of pluripotential cells, stem cells, while the functions of the tissue are performed by differentiated cells. These tissues usually show rapid renewal; such as skin, mucosae and the haemopoietic system. The rate of cell production is determined by the life-time of the mature cells.

Flexible Tissues (F-type)

These are the tissues which do not have a clear-cut hierarchical organization. In these tissues there is thought to be less of a clear cut division into primitive dividing and mature non-dividing cells. It is considered that some capacity for division is maintained even by mature cells. These tissues are generally slowly self-renewing, for example liver, kidney, lung and the CNS. However, it seems likely that purely F tissues may not exist and that, for example, the CNS is a mixed H-F tissue.

However, the different types of cellular organization imply different responses to radiation. Thus, if H-tissues are considered, while the intensity of response to radiation and the duration of recovery are dose related, the latency time between delivery of radiation and the appearance of the response is not dose related. It is dependent on the survival time of the mature cells. On the other hand the latency from dose to response in F-tissues is dose dependent. This is significant for the user of the Gamma Knife.

Tissue Architecture

Another tissue factor, beyond cellular organization is tissue architecture. The H-type tissues have a largely serial architecture, but so does the CNS. This means that function is so organized that removal of one segment can affect the whole tissue. The kidney or lung are examples of parallel organization. Thus, the destruction of many nephrons may have little effect on total kidney function, at least not until so many are destroyed, that the organ's reserve capacity is used up. This is an important concept, in that the relatively radiosensitive kidney can compensate well for radiation damage, so that it is less radio-responsive than would be expected. On the other hand, the relatively radioinsensitive CNS is radioresponsive, because damage to even a small part can produce damage to the function of the whole organ.

Cell Cycle Time (Tc) and Tumour Doubling Time (Td)

The Tc for most tumours, where it has been measured is of the order of 1 to 4 days. This is in contrast with the time it takes the tumour to double in volume; the tumour doubling time (Td). The Td for most tumours is between 1 to 2 months. This implies a considerable loss of cells during the tumour growth. Even so, in most tumours the growth rate approximates to a simple exponential function, emphasising the need for early detection and treatment.

The Fate of Irradiated Tumour Cells

Cells may be lethally injured. Lethal Injury, as stated above, is defined as a loss of capacity for indefinite proliferation. The expression of lethal injury may not be seen immediately. Usually, cell lysis occurs at the time of a subsequent mitosis, though not necessarily the first, following the radiation. In extreme cases, interphase lysis without a subsequent mitosis can occur. The timing of the observed damage may well be delayed by two additional factors. Firstly, the cells of a slowly growing tumour will take longer to die than those of a quickly dividing one. Secondly, the sensitivity of irradiated tumour cells varies with their position in the cell cycle. M phase and G_2 phase cells are more sensitive while S phase cells are more resistant. When G_1 is long cells in the early G_1 are also relatively resistant. Thus, radiation during an insensitive portion of the cycle may be associated with division delay. This division delay appears to be a linear function of dose. At all events, it is important to emphasise that the aim of conventional radiotherapy of malignant neoplasms is to destroy all neoplastic cells. Nonetheless, this may be impossible of achievement because certain tumours may require a tumour control dose which will give un-acceptable normal tissue damage. Thus, tumour growth delay is a much used clinical/radiological end-point: used to indicate that radiation gives benefit even if the tumour is not killed. The following tumour factors can influence the outcome of treatment.

The Radiosensitivity of the Tumour

This appears to be closely related to the slope of the linear component of the linear quadratic curve of cell survival, for the cells of the individual tumour.

The Volume of the Tumour

It is obvious that with a larger tumour there will be a greater number of clonogenic cells to destroy, and for a given tolerated dose of radiation, the chance of therapeutic success must be less. Moreover, there is some evidence that the clonogenic cells of larger tumours may be less radiosensitive.

Accelerated Repopulation

There is evidence that clonogenic cells that survive a dose of radiation may repopulate the tumour at a faster rate than before. This effect may be independent of tumour growth and indeed may be seen when a tumour is in fact shrinking.

The Tumour Bed Effect

This effect relates to a reduction in tumour growth or regrowth in an irradiated region and is considered to relate to damage to the surrounding tissue stroma, including the blood vessels. It is considered to contribute to retarded growth.

The Hypoxic Fraction

Those clonogenic cells that are hypoxic at the time of radiation will have a better chance of survival. This means that in the immediate period following a dose of radiation, the proportion of hypoxic cells will be greater than before the treatment.

Reoxygenation

This process, described above will contribute to increased radioresponsiveness during a course of fractionated radiotherapy. The speed of reoxygenation varies greatly from a few hours to several days.

Conclusion

The radiobiological knowledge described in this chapter has been developed in relation to standard radiotherapy. Nonetheless later chapters will show that it is also relevant for the Gamma Knife surgeon. The basic biological responses to ionising radiation are independent of the technique employed. Moreover, the linear quadratic model of cell killing is also applicable for single dose irradiation. At all events, with the degree of randomness that informs the reaction of living tissue to radiation and the capability for variability of biological response described above, it becomes obvious that quantifying the effects of treatment can be extremely difficult. This is of particular importance to the surgeon practising a young subspeciality. One of the major questions which requires an answer today is, what is the reason that not all lesions respond to treatment? Is it because of innate biological variability? In this case it is very difficult to improve the therapeutic yield. Or is this treatment failure the result of less than optimal treatment strategies? In this case research should reasonably easily lead to improved results.

Suggested Further Reading

1. McNally MJ (ed) (1989) The scientific basis of modern radiotherapy. British Institute of Radiology, London
2. Hall EJ (1988) Radiobiology for the radiologist. Lippincott, Philadelphia

3. Withers HR, Peters LJ (1980) Biological aspects of radiation therapy. In: Fletcher GH (ed) Textbook of radiotherapy. Lea & Febiger, Philadelphia, pp 103–180

5. Ionising Radiation and Clinical Practice

Introduction

This chapter will consider some of the information, outlined in the previous chapter, in relation to clinical practice in general. The concept of radioresponsiveness has been mentioned already, in relation to both tumours and normal tissues. The aim of the doctor using ionising radiation in the treatment of tumours is to kill the tumour and not damage the normal tissues. While this is desirable in theory, it is difficult to achieve in practice. In this context, the concept of the therapeutic index has been much used.

The Therapeutic Index

The therapeutic index describes the relationship between tumour damage and normal tissue damage. The tumour response curve and the normal tissue response curve are both sigmoid. This is illustrated by Fig. 5.1. The limiting factor is the appearance of unacceptable, normal tissue damage. Thus, the normal tissue damage will limit the dose. The ideal tumour is so radioresponsive that it may be destroyed without inducing significant normal tissue damage. However, the ideal tumour does not exist and in terms of the CNS, many of the tumours are markedly radioresistant. It will be noticed from Fig. 5.1 that only the lower part of the normal tissue complication frequency is charted, because to treat to higher doses would be unethical. It was at one time thought that the difference in response between normal tissues and tumours was a result of a difference in radiosensitivity between a radiosensitive tissue, the tumour, with a high rate of cell division, and a relatively less sensitive normal tissue, with a low rate of cell division. This notion was based on the concept that radiosensitivity correlated with the rate of cell division. Today, it is more conventional to consider the therapeutic index in terms of radioresponsiveness, which reflects a clinical response rather than a intrinsic biological characteristic. Other factors are thought to be more significant than the 5th 'R' of radiotherapy – radiosensitivity. Amongst these are tumour volume and the effect of fractionation in respect of the 4 'Rs'.

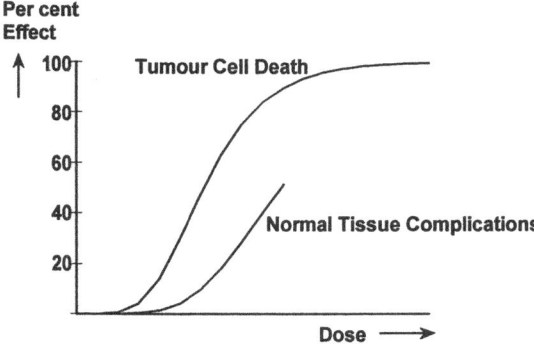

Fig. 5.1 Therapeutic Index
The therapeutic index is a measure of the difference in radioresponsiveness between the target (often a tumour) and the surrounding normal tissue. Ideally this difference should be wide, since it is the damage to normal tissue which limits the dose of permissible radiation. However, it is usually insufficiently large, so that it can be difficult to give a target dose that is large enough to destroy a tumour, because the risk for normal tissue damage would be unacceptable. It may be noted that only the lower part of the curve for damage to normal tissue is drawn in this diagram. This is to reflect the fact that a full curve would involve a risk to normal tissue of up to 100%, which would be viewed as unethical

The significance of volume on successful radiotherapy was mentioned in the previous chapter. It is a commonplace of clinical management that all attempts should be made to reduce the volume of a tumour surgically, prior to radiotherapy, in order to improve the chances of success. There is, of course, no tumour related reason why the dose should not be increased to a level which is uniformly lethal. But as stated above, the unwanted effects of radiation on normal tissues will limit the acceptable radiation dose which may be given. Thus, the rest of this chapter will consider, in a little more detail than hitherto, the effects of radiation on normal tissue, in particular in relation to clinical practice.

Time Dose Relationships

It has long been known that the biological effect of a dose is less if given in a protracted, fractionated course, instead of in one session. One of the earliest attempts to quantify this phenomenon was undertaken by Strandqvist, who tried to assess the duration of treatment on the biologically equivalent dose. Nowadays, this is called the iso-effect dose. This work, while seminal can be criticised for a number of important weaknesses. Amongst these is that it did not separate the effects of duration of dose, number of fractions and fraction dose. An even more important weakness is that the mathematical formalism used in the study, and other related formalisms derived on the same principle have subsequently been shown to be inherently incorrect. Today, the linear quadratic formula is more often used in the assessment of the effect of fractionation and duration of treatment on iso-effect doses. The effect of fractionation in sparing normal tissue is illustrated in Fig. 4.5. However, it is necessary to mention another factor, if the effect of fractionation on normal tissue is to be understood. Tissue damage may be early or late and these two categories may be distinguished as follows. There is an arbitrary cut-off between early and late complication set at 90 days. Another way of looking at this notion is to consider early complications as arising during treatment while late complications occur after treatment is over.

The effect of the duration of treatment is now considered to be complex and, at least in terms of late complications, of little relevance. Treatment duration will of course be of importance in relation to tumour response and acute reactions occurring during the course of treatment. The importance of duration of treatment is of current interest in relation to hyperfractionation and accelerated radiotherapy, but is of little more interest here. It is of limited interest in Gamma Knife treatment, for reasons which are outlined later.

The Linear Quadratic Equation and Early and Late Complications

It has been stated in chapter 4 that the α/β ratio characterised the cell survival response, for a given tissue/tumour type. A characteristic of tissues which show early complications is that they have a high α/β ratio. On the other hand tissues with late complications have a low α/β ratio. Moreover, late complications and the tissues which develop them are more sensitive to fractionation, while early complications and the tissues which develop them are less sensitive to fractionation. In general, the α/β ratio for tissues showing early reactions is 7–20 Gy and for tissues showing late reactions it is 0.5–6 Gy.

Early and Late Reactions

Early Reactions

As stated above, these reactions occur while radiation treatment is still going on and/or before 90 days after the start of treatment. They are characterised by a high α/β ratio. They occur chiefly in H-tissues, are due to an effect on parenchymal stem cells and are largely reversible. Their intensity is dose dependant and the time taken to resolve is dose dependent, but the latency from start of treatment to appearance of complication is not dose dependent. It is related instead to the life span of the mature cells in the tissue concerned. These complications are not fractionation sensitive and for a given dose, on the whole, the shorter the duration of treatment, the greater the injury. Typical early reactions are the erythema and desquamation seen in the skin.

Late Reactions

These reactions have a low α/β ratio. They occur in all sorts of tissues and are thought to be related to effects on connective tissue rather than parenchyma. They occur after 90 days and often after many months or even years. At all events they occur after the cessation of treatment. The intensity of response and the shortness of the latency to the complications' inception are dose dependant. They are in general irreversible. They are fractionation sensitive but not sensitive to the duration of treatment. Subcutaneous fibrosis, skin telangiectasia or radiation damage to the CNS are all late reactions.

Radiation Quality

There is another factor which plays a part, if all be it a minor one for the present purpose; the quality of the radiation. The iso-effect doses, described so far relate to the differences in biological effect of a given dose, when that dose is given over varying periods and in varying fractions. **R**elative **B**iological **E**ffect (RBE) relates to the difference between biological effect, for the same dose, when different forms of radiation are used. The study of the RBE is made easier by using the effect of a given dose of ^{60}Co as a standard and comparing other forms of radiation with this standard.

Conclusion

The present chapter has outlined some general principles related to the effect of radiation on living tissue. It has been shown that dose, while important, is not alone the only relevant factor. The use of the linear quadratic formula has made the assessment of different factors somewhat easier to assess and even predict. Moreover, it is important to appreciate that it is the reaction of normal tissues to radiation that limits the dose that may be given in the treatment of tumours by radiotherapy. Finally, the application of these various factors to the specifics of Gamma Knife treatment, will be considered in chapter 7.

Suggested Further Reading

1. McNally MJ (ed) (1989) The scientific basis of modern radiotherapy. British Institute of Radiology, London
2. Thames HD, Hendry JH (1987) Fractionation in human radiotherapy. Taylor and Francis, London

6. Development of the Gamma Knife

Introduction

The first Gamma Knife was constructed after more than two decades of research, in both the laboratory and the clinic. The primary intention, at the start of this research was as indicated in chapter 1, non-invasive treatment, rather than a commitment to a particular technology. Ultrasound was considered and rejected, because at that time it could not be used with precision without opening the skull. Leksell's first paper with the word radiosurgery in the title was published in 1951. Thus, the concept of a non-invasive surgical procedure and the development of a simple usable stereotactic system seem to have been contemporary. Moreover, it is easy to forget today, when the major indications for Gamma Knife surgery are tumours and malformations, that the limits of imaging techniques applying to open stereotaxy also apply to radiosurgery. In the nineteen fifties and sixties radiological tumour delineation was approximate, seen in relation to the precision of stereotactic treatment. Thus, the early experimental work was performed and the first clinical instruments were designed with a view to making cerebral lesions, for treating functional disorders. It seems likely that it was considered particularly important to avoid the potential dangers of surgery when treating conditions which did not present a short term threat to life.

Preparatory Basic Research

The first steps in the basic research were taken in association with the distinguished radiobiologist, Börje Larsson, at the University of Uppsala, at that time a very young man. The first matter to be decided was the choice of a suitable type of radiation. The radiation should be able to penetrate to the desired depth, without delivering potentially damaging energy to the tissues along its path, in other words radiation with a low Linear Energy Transfer or LET. Furthermore, the method of delivery of the radiation had to fulfil certain criteria. It must be possible to localize the radiation target accurately. Moreover, it should enable the design of a radiation field with a high dose where it was needed, at the target and a very sharp dose fall at the edge of the target, enabling protection of tissue around the target from damaging radiation. High energy protons from the 185 MeV synchrocyclotron, at the University of Uppsala, close to Stockholm provided such a source of radiation. The localization was achieved by fitting a Leksell stereotactic frame to the head and directing the beam in accordance with the frame's axes. The desired dose distribution could be achieved by cross firing a number (20 to 22) of small diameter beams. It should be noted that with this cross firing technique, deceleration of the protons, producing a Bragg Peak was not the aim. Using high energy proton beams, with the low LET characteristic of protons, which have not yet begun to decelerate, cross firing produced a very low radiation-tissue interaction along the individual beams, with a concentration of radiation activity in the volume where the beams crossed (Fig. 6.1). A consequence of this design is that the effects of the technique cannot be compared with the effects of Bragg Peak radiation (see Fig. 3.5), since the two types of lesion are radiobiologically dissimilar. It must be

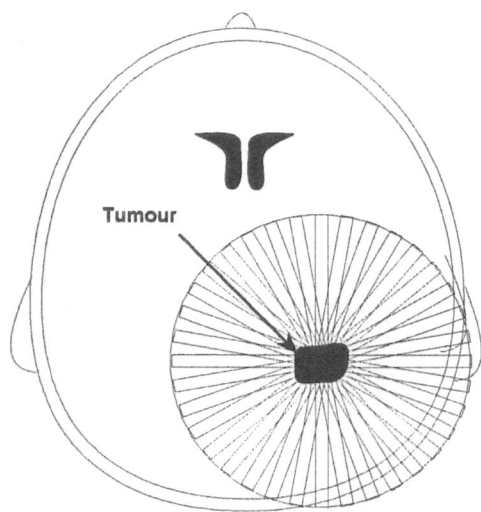

Fig. 6.1 Focal Radiation (Cross Firing)
Focused radiation damage produced by the Bragg peak was illustrated in Fig. 3.5. The other commonly used way of producing focused irradiation volumes is to crossfire radiation at a given target. This is illustrated here, though only the part of the radiation nearest to the target is shown for the sake of clarity. This illustration is comparable with Fig. 3.5, where the same lesion is surrounded by radiation using the Bragg Peak method

emphasised that it is the narrow diameter of the beams which enables the great precision of the radiation field (Fig. 6.2).

Since the intended clinical application was a lesion in the brain parenchyma, the early experiments studied the characteristics of lesions in the cerebrum of the goat. Goats were used because their skulls are large and rigid enough to allow the application of a human

40 mm diameter Collimator

8mm diameter Collimator

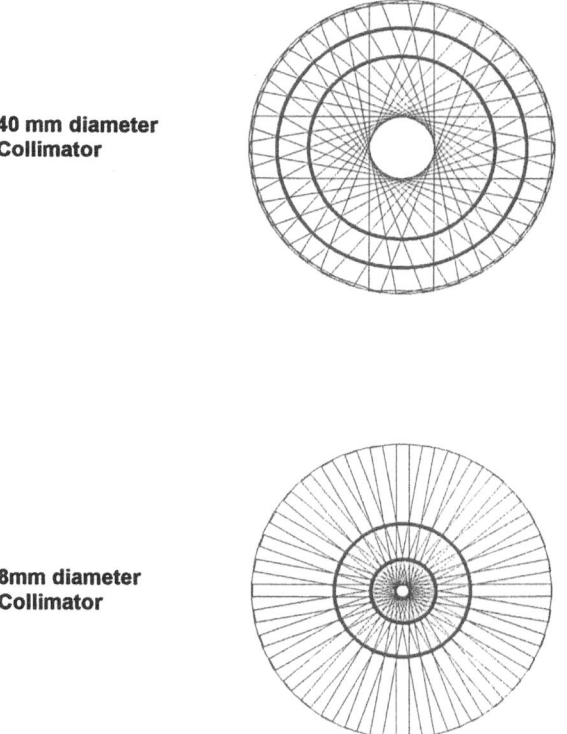

Fig. 6.2 Effect of Collimation Width on Dose Precision
This illustration shows the distribution of the lines joining equivalent points of intersection, between the beams of radiation arranged around a central target. In the upper diagram the beams are broad while in the lower they are much narrower. It is obvious that with narrow beams, in the lower picture the radiation is very concentrated at the centre of the irradiated volume, with a very sharp fall in dose away from the high central dose. In the upper diagram the fall off is more diffuse and spread. Thus the diagram indicates the importance of many very narrow radiation fields for producing a well defined radiation volume, with a sharp dose at the edge. The dose along each individual beam is of course low and has little effect. This is the situation in the Gamma Knife

stereotactic frame. Moreover, their brains are large enough, in proportion to the lesion to avoid such complications as generalised radiation oedema.

Disk shaped lesions were produced, because the disk shape would be most appropriate when performing a thalamotomy, in patients. 200 Gy was used and a somewhat stereotyped lesion was produced. After four weeks, necrosis, degeneration and inflammatory reactions could be observed. Over the subsequent weeks there was intense cellular activity, characterised by phagocytosis and the beginning of scar formation. After a year, a stable glial scar had developed. Like the radiation volume, the lesions were disk shaped. The lesions were also characterised by a **sharply defined edge** lying at the region of maximum dose fall, roughly equivalent to the 50 % isodose curve. Moreover, there was very little increase in lesion volume over time. If the dose was doubled to 400 Gy a much more marked and diffuse lesion was seen, with even a risk of oedema in the entire cerebral hemisphere. These lesions were chronically inflammatory and did not result in stable avascular scars. Thus, there was clearly an upper acceptable target dose that should not be exceeded. Another finding was that above a certain threshold, of about 150 Gy, the time to the development of a lesion was within limits a function of dose. The existence of the threshold indicated that there was also a lower dose limit, below which no adequate lesion would develop. All these findings indicated that in radiosurgery, the dose would be critical.

The First Patients

The above is only the merest outline of the detailed research which was performed before the first patients were treated radiosurgically. However, it indicated that sharply defined small radiosurgical lesions, appropriate for the treatment of functional disorders could be relatively easily obtained. The next step was a clinical trial. It was still too early to build a new dedicated machine for the first patients so they were treated using already existing machines. The first patient reported following radiosurgery – in 1955 – suffered from long lasting, severe chronic schizophrenia. She was treated with 280 kV X-rays from an industrial X-ray tube. 40 Gy were delivered first to the right rostral internal capsule, followed 33 days later by the same dose to the left. The result was indeterminate but the following conclusions could be drawn. After two years follow-up, the patient had suffered no harmful effects of the treatment. She responded to ECT and chlorpromazine afterwards, something she had not done before. She was eventually able to return home a year later, after a 4 year stay in a mental institution. In retrospect, the dose today seems rather low, but the result was sufficiently encouraging to stimulate Leksell and his colleagues to continue with the method.

The next patients to be treated, suffering from Parkinsonism or chronic pain, were treated using the Uppsala synchrocyclotron. The results of this series were also promising enough to suggest that radiosurgery had a definite future, as a clinical technique. However, Leksell felt that the synchrocyclotron, using the cross firing technique, was too inconvenient for routine clinical use, so an alternative technology was required. A number of options were considered, including accelerated ions or electrons, supervoltage X-rays, neutrons and gamma emitting nuclides. The final choice fell on gamma nuclides, because the radiobiological characteristics of this form of radiation were so similar to those of the protons that had been used hitherto. Of available nuclides, ^{60}Co was finally chosen, not least because its half life of 5.27 years was seen to be the most practical.

The First Gamma Unit

Requirements

The requirements of the instrument, the main application of which was to be functional neurosurgery, were as follows.

1. Adequate **precision.** This requires both accuracy of localization, accuracy of the radiation field and immobility of the patient, in respect of the radiation. The accuracy of *localization* was achieved with the stereotactic frame and high quality X-rays taken with the central beam at right angles to the frame, as described in chapter 2. The accurate *radiation field* was achieved by beaming 179 cobalt beams, mounted on a hemisphere towards a sharply defined focus at the centre of the hemisphere. The beams were directed through collimators with a rectangular cross section in order to obtain the desired disc shaped lesion. Two collimator sizes were available, 3×5 mm and 3×7 mm. The collimators were held in a helmet. Immobility of the patient was achieved by rigid fixation of the stereotactic frame attached to the patient's head, to axis rods attached to this helmet. The machine incorporated a hydraulic system to move the target to the focus of the machine. The use of the same frame for X-rays and treatment resulted in concordance of the stereotactic target on the X-rays and in the machine. The extreme rigidity of the head fixation and the lack of moving parts during radiation are central to the high degree of accuracy of the technique.

2. Acceptable **treatment time.** It was clearly desirable that the treatment should be carried out as quickly as possible. It is not easy for often elderly patients, with severe pain or paralysis agitans to remain in one position for long periods. The major determinant of the treatment time is the energy of the radiation source The ^{60}Co used provided a beam of 1.17 or 1.33 MeV. Leksell commented in his description of the first cases that the treatment time was on the long side and discussed the possibility of using isotopes with a higher energy in the future.

3. The **brain integral dose** should not be greater than that commonly accepted for a single radiation of malignant brain tumours. With the tiny lesions produced with the small collimators used, the geometry of the machine ensured that the spread of radiation dose was very restricted, even with a maximum dose of 200 Gy.

4. Adequate **radiation protection** for the patient's body and the treatment personnel. The patient's protection is related to a single exposure and was guaranteed by the geometry of the machine, which precluded the spread of any unacceptable amount of radiation away from the focus. For the personnel who would be repeatedly exposed to the machine, their absence from the treatment room during treatment and the massive casings and doors of the device were the major factors. Of course, the usual routines for any department using radioactive material, of repeated measurements of radiation in the locale, radiation absorption badges and the like were employed.

Early Experience

The first patients were treated after installation of the prototype machine, in the Sofiahemmet hospital in Stockholm, early in 1968. This protype was called a Gamma Unit and the first patients were treated with gamma thalamotomy for intractable cancer pain. 160 to 250 Gy were used. This was a clinical material which inevitably allowed post-mortem examination of the lesion produced, only a relatively short time after the treatment. The lesions and the dose time relationships of the lesions were very similar to those seen in the experimental work on animals, using the synchrocyclotron. However, the very first patient ever treated was a tumour case, a small craniopharyngioma. This patient was treated in the autumn of 1967 before the machine had left the factory. Other diagnoses treated in the early days were trigeminal neuralgia, Parkinson's Disease, small arteriovenous malformations, acoustic neurinomas and pituitary adenomas.

As stated above, the first Gamma Unit had been devised for producing disc shaped lesions appropriate for functional work. With the advent of dopamine, at the beginning of the 1970s, the number of Parkinson patient referrals decreased. At the same time, the widening of the indications for Gamma Knife surgery to include tumours made the disc shape less than optimal. Moreover, the collimators were too small for most tumours and malformations which might be reasonably treated by the method. Thus, it was deemed necessary to design a second Gamma Unit.

The Second Gamma Unit

In this unit 179 beams were obtained from 179 ^{60}Co sources, as in first Gamma Unit, but in this new machine the collimators were circular in cross section and described as either 8 or 14 mm in diameter. These distances referred to the diameter of the 50% isodose. The larger and more spherical dose distributions, available with the newer collimators were more appropriate for the treatment of malformations and tumours. Indeed, the machine was constructed to accommodate a collimator size range from 4 to 28 mm. Subsequently, 4 mm collimators became available. Larger sizes were not used because the dose fall becomes more gradual with larger collimator sizes. This matter is dealt with in more detail in a later chapter. The new computer programme available with the new unit permitted the calculation of multishot dose-plans, for any of the collimator sizes in the above-mentioned range. This greatly increased the precision with which a dose distribution could be tailored to fit the shape of the lesion that was being treated. Usually, the aim was to fit the 50% isodose to the edge of the lesion. Figure 6.3 shows a diagram of a longitudinal section through this version of the Gamma Unit. This shows all the necessary principles. Subsequent models have only changed in matters of detail not of principle.

The Third Gamma Unit

This exists in two models and is called the Gamma Knife. Both models have 201 sources and four collimator sizes 4, 8, 14, and 18mm as routine. The first model has basically the same design as the second unit but with more sources and a slightly larger helmet. The second model is somewhat simpler with the opening in the helmet at a simple right angle to the long axis of the bed of the machine. This simplifies the movement of the bed taking the patient in to the inside of the

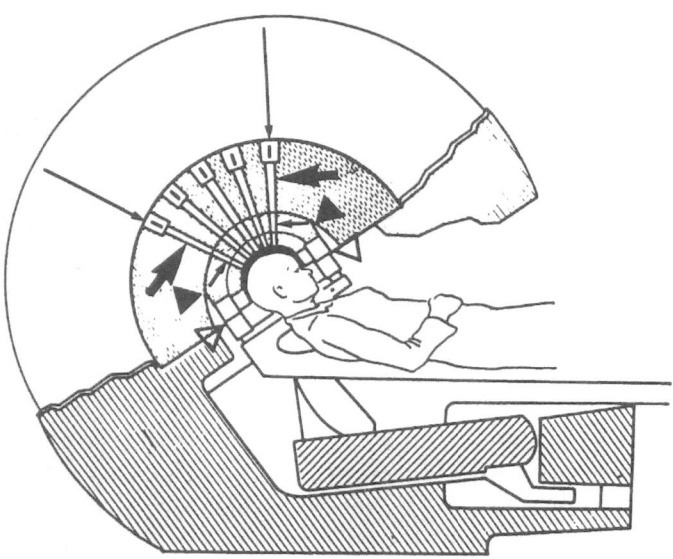

Fig. 6.3 Gamma Unit Diagram

This shows a longitudinal section through a Gamma Unit, with the patient in place. Note the relaxed patient position. The head is fixed to the inner helmet (filled arrow-heads), by means of attachments which are fixed to the frame and adjusted to bring the target to the focal point of the radiation. These are not visible in this diagram. The inner helmet is fixed to the bed (open arrow-heads) of the instrument and moves into the machine to take up the position indicated in this diagram. There are also radiation insulating doors, at the opening of the machine, not shown in this diagram, which must open prior to the movement of the bed into the machine. In the position shown, the helmet, with its collimators (small arrows) are placed exactly coaxial with the channels – 5 of which are shown (broad arrows) – drilled in the outer helmet, which is mounted within the machine's interior. At the distal end of these channels the ^{60}Co is placed (long thin arrows). Thus, as the radiation is effectively shielded by the two helmets, until the inner helmet is in position, the dose may be accurately calculated to begin at the moment the inner helmet is in position. The only movement in the system is that of the bed which draws the patient in and out, and the doors at the mouth of the machine which open prior to the movement of the bed. Thus the extreme simplicity of the design contributes to great simplicity in use and reproducibility. When the treatment is over, the bed automatically moves out and the doors close

instrument. Thus, in this model the patient is only moved horizontally into the machine and not in and up as in previous models. Again, the helmet is slightly larger than in the earlier models.

Summary of the Current Gamma Knife Method

Firstly the frame is applied to the head. This frame functions as otherwise in stereotactic procedures as both an axis system and a platform. The axes work as previously described. It is a platform for X-ray indicators to secure accuracy of measurement of the axes. Moreover, the frame is a platform to which adapters to fix it to the helmet of the Gamma Knife may be attached. These adapters, fixed to the same frame, in a position that has remained unchanged since the dose-planning X-rays were taken ensures that the axis system recorded on the X-rays and the axis system in the Gamma Knife are exactly concordant.

Nonetheless, there is a practical problem relating to the application of the frame for use with the Gamma Knife, which does not apply for open stereotactic surgery. In this latter situation, instruments can gain access to any point within the cranium. This is not the case with the Gamma Knife. The size of the helmet restricts the degree of movement of the head with its attached frame so that targets with an eccentric intracranial location require special frame placement. The aim is to place the target as near the centre of the frame as possible. This is considered in more depth in chapter 9.

The frequency of peripheral Gamma Knife targets has increased, since metastases have become a common indication for this form of treatment. Thus, in this situation it may be a good idea to have a trial placement of the frame, the day before treatment. No fixation is performed, but the frame is assembled in an advantageous way and notes are made as to the requirement for rotation of the frame, if necessary. It must be admitted that in the early days it was thought that a peripheral lesion location could represent a contraindication for the Gamma Knife. Firstly, it was considered that they would be difficult to fit into the machine. This is almost invariably not true (see chapter 9). Secondly, they were more accessible to conventional surgery. This was a more serious consideration. With modern experience and the widespread use of the machine, it is no longer, of itself, a contra-indication to the use of Gamma Knife surgery. Thirdly, it was thought that a peripheral lesion would be associated with cortical damage, and therefore an increased number of complications. There is, at the time of writing, no published evidence to support this view.

While it is true that a peripheral lesion placement is not of itself a contraindication to treatment, it does carry further implications than the potential problems described in the previous paragraph. With the increasing use of MRI as the localizing technique, a peripheral location can be a problem in relation to the nature of MRI, as it is practised today. There is, on MRI images some degree of anatomical distortion, which increases with increasing distance from the centre of the film. Thus, this is another reason for trying to place the target as near the centre of the frame as possible.

Thus, all this boils down to is that for a variety of reasons it is advantageous to place the frame with the lesion as central as possible. This means that frame application is somewhat more flexible and on occasion a good deal more difficult than the application for open stereotactic procedures. Another group of difficulties relating to frame placement arises during the treatment of small arteriovenous malformations, which may be hidden by parent vessels or overlying bones in conventional projections. In this situation, rotation of the frame can be most helpful.

Finally, the use of computerised images to define a target *point,* for conventional open stereotactic procedures is not quite the same as the definition of a target *volume,* for Gamma Knife treatment. The definition of such a volume places more stringent demands on the specialist, responsible for deciding the extent of the target volume, prior to dose-planning

Dose-Planning

The actual dose-planning is considered outside the scope of this book. Suffice it to say that by the manipulation of a single shot or a combination of shots, the surgeon attempts to produce a radiation field with the highest possible dose to the target and the lowest possible dose to the surrounding brain tissue. The constraints on the surgeon will in fact be discussed in some more detail in relation to the individual pathological processes, which are the subject for treatment in the Gamma Knife.

Suggested Further Reading

1. Arndt J, Backlund E-O, Larsson B, Leksell L, Norén G, Rosander K, Rähn T, Sarby B, Steiner L, Wennerstrand J (1979) Stereotactic irradiation of intracranial structures: physical and biological considerations. INSERM Symposium 12: 81–92
2. Backlund E-O (1979) Stereotactic radiosurgery in intracranial tumours and vacular malformations. In: Krayenbühl H et al (eds) Advances and technical standards in neurosurgery, vol 6. Springer, Wien New York, pp 1–37
3. Backlund E-O (1992) The history and development of radiosurgery. In: Proc. International Symposium on Radiosurgery (Pittsburgh). Elsevier, New York, pp 3–10

4. Larsson B, Lidén K, Sarby B (1974) Irradiation of small structures through the intact skull. Acta Radiol 3: 512–534
5. Larsson B, Leksell L, Rexed R, Sourander P, Mair W, Anderssom B (1958) The high-energy proton beam as a neurosurgical tool. Nature 182: 1222–1223
6. Leksell L (1983) Stereotactic radiosurgery. J Neurol Neurosurg Psychiatry 46: 797–803
7. Leksell L (1971) Stereotaxis and radiosurgery. An operative system. Ch C Thomas, Springfield, IL
8. Leksell L, Herner T, Lidén K (1955) Stereotaxic radiosurgery of the brain. Kungl Fysiografiska Sällskapets i Lund Förhandlingar 25: 3–10
9. Leksell L, Larsson B, Andersson B, Rexed B, Sourander P, Mair W (1966) Lesions in the depth of the brain produced by a beam of high energy protons. Acta Radiol 54: 251–264

7. Radiophysics, Radiobiology and the Gamma Knife

Introduction

The purpose of this chapter is to relate the general knowledge, outlined in chapters 3, 4 and 5, to the particular conditions pertaining to Gamma Knife radiosurgery. This is a fascinating area, where new methods and findings are expanding and enhancing our knowledge, at an increasing rate. One particularly important new aspect that will be mentioned is research concerned primarily with the effects of single shot radiation on tissue. It has been mentioned (see chapter 1) that the term radiosurgery has unhappily become somewhat controversial. Nonetheless, all concerned with treatment techniques, which may loosely be said to fall under the generic term radiosurgery agree that extremely precisely localized, single treatment session radiation are two of the hallmarks of the techniques.

Radiophysics and the Gamma Knife

Energy

Gamma radiation is of course non-particulate electromagnetic radiation, to be considered either as waves or photons. The gamma radiation produced by ^{60}Co has two energies, reflecting two distinct radioactive breakdown pathways. The gamma radiation from these two reaction series has an energy of either 1.17 or 1.33 MeV depending on which radioactive breakdown pathway is being considered. With radiation energy within this range, according to the description in chapter 3, most of the interaction between radiation and irradiated tissue can be expected to be mediated by Compton scattering and to a lesser degree pair production. The energy level of this radiation is sufficient to give it a high power of penetration. It has a low LET (see chapter 3). The narrow beams, essential to the technique, are produced by a construction which forces the radiation through collimators in the form of small metal tubes, mounted in the machine's helmet. The size of the collimators is defined in terms of the diameter of the 50 % isodose around the centre of the target. These collimator sizes are 4 mm, 8 mm, 14 mm and 18 mm.

Radiobiology and the Gamma Knife

Factors Affecting Cell Survival

It must be clear that those elements of radiobiology related to fractionation are not relevant with a single session treatment. Thus, the 4 'Rs' of conventional fractionated radiotherapy are of less importance. The place of any cell in the cell cycle will be entirely random. By the same token reassortment will play no part in this sort of treatment. Repair and repopulation will not occur during a treatment which takes under an hour. Neither will reoxygenation. On the other hand, oxygenation at the time of radiation can play a part. Moreover, the very high single dose should be effective in reducing the chance of repair and repopulation after cessation of treatment. The effect must be achieved by a combination of single lesions and the summation of potentially lethal hits, in keeping with the linear quadratic concept. Factors related to the therapeutic index are considered a little later.

Dose

As the individual beams are characterised by a low LET, little energy (or dose) is absorbed along their individual paths. However, where the 201 narrow beams meet, at the focal point of the Gamma Knife, there is a

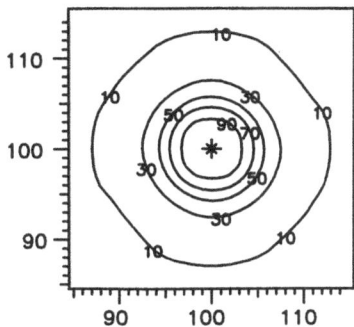

Fig. 7.1 Gamma Knife Isodose Curves
Cross section of isodose lines in the X-Y (axial) plane. It may be seen that there is a zone of rapid dose fall, between the *90%* and *50%* isodose lines which is beginning to slacken at the *30%* isodose and is much slacker beyond this isodose line

summation of dose, which gives a high **target dose.** The geometry of the machine, with the large number of narrow beams, arranged around the surface of a sphere enables the construction of very precise dose distributions, with a very sharp fall in dose at the 50 % isodose. It is for this reason that the 50 % isodose is made to conform with the edge of any lesion to be treated. Figure 7.1 shows a two dimensional representation of the isodose lines in the X-Y plane. Figure 7.2 shows the relative dose fall as a function of the distance from the central maximum dose. The very sharp dose fall, between 80 % and 30 %, a result of the machine's geometry as stated above, enables the delivery of a high target dose to the lesion, within the 50 % isodose and a low **brain integral dose** to the surrounding brain. It is necessary to determine a minimum acceptable target dose and place this at the edge of the lesion, at the 50% isodose. The appropriate dose for different types of lesion varies for each lesion and will be discussed subsequently. Thus, it is the geometry of the machine rather than the technique of fractionation, which spares normal tissue with the Gamma Knife.

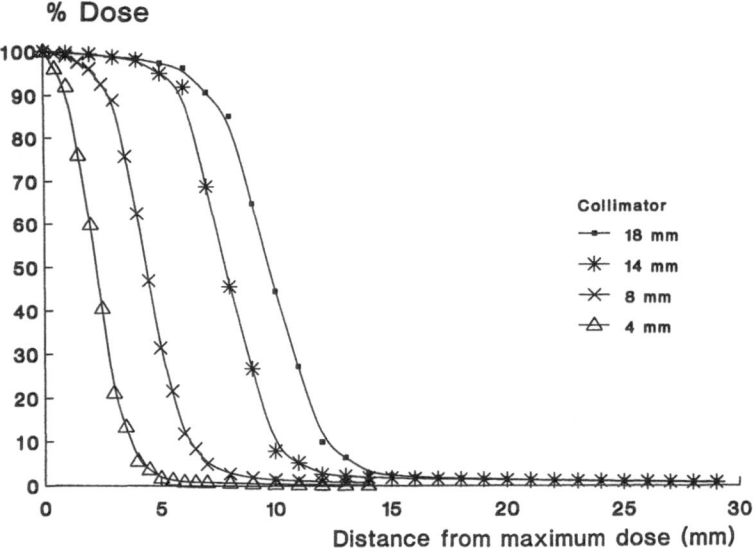

Fig. 7.2 Gamma Knife Relative Dose Fall
Graph of the relative dose fall as a function of distance from the central maximum dose, for the four collimators. Note again the very rapid dose fall between 80% and 50%. Note also that the sharpest dose fall, and thus the most accurate application of the rapid dose fall region at the edge of the target applies to the smaller collimators

Dose Rate

The half life of ^{60}Co is 5.27 years. This means that with the passage of time, the dose rate for the Gamma Knife is steadily decreasing. Or put another way, the same treatment with the same dose takes longer and longer the older the machine becomes. There is some evidence that reducing the dose rate reduces the biological effectiveness of radiation especially if the dose is less than 1 Gy per minute. Doses as low as this are to be found at the lesion edge, with even a fairly new Gamma Knife, though their significance is not clear at this time. However, as the results of Gamma Knife surgery are so good it seems unlikely that low dose rate within the target is of any great importance.

One advantage of the geometry of the Gamma Knife is that since a lower isodose is associated with a lower dose rate a measure of protection is provided for the normal tissue outside the target volume. This is because the maximum dose rate is set by the activity level of the gamma source. Thus, for a single shot, the dose rate at the 50% isodose will be half that of the dose centre and the dose rate further out from the centre will decrease in proportion. This means that with a dose of 100 Gy at the centre, with a 3 Gy/min dose rate, the dose rate at the 50% isodose will be 1.5 Gy/min and at the 10% isodose 0.3 Gy/min, which is a dose rate low enough to reduce the effectiveness of the dose, thereby producing a measure of extra protection in the surrounding tissue.

Dose Volume

This is a factor under close scrutiny at the present time. This has been mostly concerned with the treatment of arteriovenous malformations and will be mentioned again during the chapter related to that topic. In general, the effect of lesion volume on therapeutic success in controlling the lesion should be considered to be the same as for radiotherapy. A larger lesion has a larger number of clonogenic cells to control, which for a given dose will make control less likely. It is logical to consider the effect of treatment volume on the surrounding tissue at this point. Van der Kogel found that correlating the length of rat spinal cord irradiated with the development of radiation induced paralysis showed that there was a relationship between these variables. Briefly, if the length of irradiated cord were 2 mm or less, no white matter necrosis developed, even with a single dose of more than 80 Gy. On the other hand for a length of 2 cm or more, 20 Gy consistently produced white matter necrosis. Between these two limits, there was a gradual increase in the occurrence of paralysis with increasing length of cord irradiated. One very relevant finding, consistent with these observations is that there is as yet no patient reported to have a facial palsy following treatment of an acoustic neurinoma, with a volume of less than 1000 mm^3.

These findings are important and in keeping with clinical experience. It is also important to be clear about the fact that an increase in target volume automatically increases the volume of normal tissue receiving a significant dose of radiation; or in other words increases the integral dose to the normal brain. This is the factor which limits the size of lesion which can be treated by the Gamma Knife. The technique is fully capable of producing much larger radiation fields than are currently in use. So, it is not the machine design which limits the target volume, which is appropriate for radiosurgery. It is the decreasing tolerance of normal brain with large integral doses.

Dose Homogeneity

The Gamma Knife radiation field is inherently inhomogeneous. This must be so, if the radiation dose is high and is distributed between a maximum and 50% of that maximum, across any given target. Clearly this does not matter all that much, since the technique works. However, it is untraditional for teletherapy, where in conventional radiotherapy it is usual to strive for as homogeneous field as possible. At the present time, it is unclear if Gamma Knife results could be improved by aiming at more homogeneous fields than are currently in use. It is an issue which is being currently examined within the Gamma Knife milieu.

Delayed Radiation Damage of Nervous Tissue and the Gamma Knife

Inevitably any discussion of radiobiology must consider the particular characteristics of the tumour being irradiated and the characteristics of the tissue surrounding the tumour, or from which it grows. Some of the more specific biological tumour related topics will be mentioned in chapters related to individual tumours. However, there are some general considerations relating to tumours and it is clearly appropriate to discuss the radiobiological qualities of the normal tissue surrounding intracranial tumours, the tissues of the central nervous system.

Radiation Tolerance of the CNS

The effect of single shot radiation on the nervous system has received considerable attention, in particular from Professors Börje Larsson of Uppsala Sweden, now in Zurich, Switzerland and Albert Van der Kogel of Nijmegen Holland. What is presented in the following paragraphs relies heavily on the writings of these two experts.

Using a variety of mainly small experimental animals, it would seem that the following tentative conclusions may be drawn. Irradiation of central nervous tissue, either brain or spinal cord, produces damage which appears to be dose related and affects either or both the white matter myelin and the vascular endothelium. These changes are thought to be due to damage to the endothelial or glial cells, in particular oligodendroglia, which have a relatively high rate of cell division, in a tissue with an otherwise low rate of cell division. These processes appear to be interrelated in some ways, though the details of this interrelationship is not clear. There appears to be a threshold dose involved in the production of CNS necrosis, in these animal models and this was of the order of 20 Gy, for a single radiation dose, though the relevance of this for clinical practice is not at present defined.

Dose Latency Relationships

What sort of tissue reactions to radiation are seen in the CNS? Are they early or late? This is not entirely a straightforward matter. An early normal tissue reaction (within a week or two after treatment), consisting of reversible oedema has been described, but this is exceptional and will not be referred to again. According to the characteristics of early and late radiation complications, outlined in chapter 5, the tissue reactions described in the preceding paragraph are late reactions. Nonetheless, it is just these reactions that are the basis of the lesion in thalamotomy, described in chapter 6 and these lesions occur before 90 days. According to one of the accepted usages of the terms early and late, in this context, that would make the complications early. The other definition of early and late complications, related to duration of treatment is obviously not relevant, in respect of the Gamma Knife, with its total dose in a single session technique. All complications occur after cessation of treatment. On the other hand, in traditional, fractionated, radio-therapy a striking feature of the development of clinical CNS damage, following a total dose of the order of 50 Gy is the long latent interval from radiation to necrosis. The explanation of these apparent discrepancies of definition is that the latent interval is dose dependant and the terms early and late were defined in relation to the doses used in conventional radiotherapy, not those used in functional radiosurgery. Moreover, as recounted in chapter 5, this is to be expected, since the latency of late reactions is dose dependent for 'F' tissues, of which the CNS, at least to some extent is one. And indeed, a complication latency of many months is also seen with the Gamma Knife, with the lower doses used in the treatment of space occupying lesions.

The notions described in the previous paragraph are supported by a number of observations. Experimental work, on rat spinal cord has shown that there is a general tendency for the latency from radiation to CNS damage to decrease, with increasing dose, by a factor of the order of 2 days/ Gy. However, the relationship is not a simple one. Such a finding is in keeping with the earlier observations, mentioned in chapter 6, concerning the speed of the development of lesions, following Gamma Knife and cross-fired proton beam therapy being a factor of dose; within certain threshold limits. Thus, to produce a CNS lesion, appropriate for thalamotomy (2-5 mm diameter), within 1–2 weeks, a dose of 200–300 Gy was required. Lower doses produced the effect with a longer latency, down to a threshold of about 130 to 150 Gy. Below this level no lesions were noted. Using the same size of lesion, with dose levels above 400 Gy, there was a risk of hemisphere swelling, indicating a quite different set of biological processes at this dose level.

This brings us to another matter; the protection of the brain outside the target volume. This is achieved by designing the Gamma Knife to take advantage of the radiobiological characteristics, just described. It has been mentioned earlier that the 50 % isodose was placed at the edge of lesions produced with cross-firing techniques, whether proton beam or gamma ray. The reason for this was to limit the late spread of the lesion on the basis of two ideas. Firstly, the latency of the white matter necrotic effect was dose dependent. With a rapid dose fall there should also be a rapid rise in necrosis latency, concomitant upon the falling dose. Secondly, there was a lower threshold for white matter necrosis, below which no necrosis would occur. Thus, the smaller the volume enclosed within this threshold dose, the larger the volume outside this dose which could not be damaged. It is emphasised that the production of a thalamotomy or other destructive lesion in the brain, with the Gamma Knife is a conceptual departure for therapeutic radiation techniques. Here, the aim is to damage the normal tissue. Thus, any complications, as well as the therapeutic effect are related to damage inflicted on one and the same tissue. The sharp dose fall at the target edge, together with lesion forming dose levels within the target volume are the mechanisms by which the normal tissues are protected.

In the previous paragraphs the reactions to be expected in an F tissue are discussed. However, the brain is considered a mixed tissue with H elements and F elements. It is known that some of the glial cells of the normal brain are refurbished from a layer of subependymal stem cells. However, the contribution of the H component of the brain to its reactions to radiation are not clear at the time of writing.

CNS Radiation Tolerance and Dose Volume

See above under Radiobiology and the Gamma Knife (page 69).

CNS Radiation Tolerance and Fractionation

As the CNS behaves largely as an F-tissue, showing late complications, it is to be expected that it is sensitive to fractionation. This is indeed the case. On the other hand the significance of the duration of treatment is not at all clear. While fractionation sensitivity is an advantage which the Gamma Knife surgeon has hitherto not used, this does not matter, since the advantage is related to preservation of normal tissue. This the Gamma Knife surgeon achieves through the accuracy and geometry of his instrument, as described above. Thus, single dose treatment strategies remain the standard for Gamma Knife surgery, as it is practised today. However, there is a debate within the milieu about the potential benefits of doses in two fractions for larger lesions. The importance of this remains to be seen. The author allows himself the reflection that there may well be situations where fractionated treatment has advantages. Some would say that this then makes the treatment a form of radiotherapy and not radiosurgery. However, staged surgical treatment of difficult lesions was employed by Cushing for meningiomas, at the birth of modern neurosurgery. It is still an accepted option today, since it has been recently advocated by Spetzler, in the management of difficult arteriovenous malformations. Thus, for most relevant lesions, the Gamma Knife remains essentially a surgical tool, with its accurate application being dependent on surgical competence, irrespective of the number of treatment sessions. Nonetheless, while every physician should wish to improve his technique, the Gamma Knife's track record, when used as a single session radiosurgical tool is not lightly to be disregarded.

Tumour Tissue and the Gamma Knife

All that has been mentioned so far in this chapter applies to the reactions of normal nervous tissue, nothing has been said about the reactions of the pathological tissue. In other words, single tissue radiation has been considered, rather than the more conventional arrangement of a pathological target, surrounded by a normal tissue at risk. Firstly, since the Gamma Knife has recently been applied to the treatment of malignant targets the tumour factors relating to such targets may be considered first. This is because the aim of treatment is identical with that of conventional radiotherapy. This aim is sterilization and disappearance of the tumour. This will require a very high dose and a higher rate of complications is acceptable, due to the poor prognosis of the condition being treated. Perhaps the most significant characteristic predicting success is tumour volume. It would appear that metastases, which are excellent lesions for Gamma Knife surgery are less susceptible if they are over 3 cm in diameter. The significance of volume is even more apparent in relation to glial tumours, whose visible volume is usually larger than this, and whose real volume is even bigger.

The technique allows a radiation dose, in one session, which may be the biological equivalent of the whole dose of radiotherapy. It may be even more effective. Again, it is the rapid fall of dose at the edge of the lesion which permits adequate single session treatment. Other factors related to the effect of treatment, such as oxygenation, tumour bed effect, and accelerated repopulation are of little interest with the Gamma Knife.

The aim of Gamma Knife treatment with benign tumours may be slightly different from that described in the previous paragraph. Firstly, it has been widely accepted that growth arrest without disappearance is an acceptable result. This has been applied for acoustic neurinomas, meningiomas and pituitary adenomas. For such lesions, in particular where alternative therapies are available, it is imperative that there are a minimum of normal tissue complications. This restricts the doses that may be employed. Though let it be said, this has not hindered excellent results with the technique. Recently, Larsson has argued that the desired differential radiosensitivity requirement for the treatment of cancers, with the implication of death to all malignant cells may not be a requirement for life-long tumour control of benign intracranial tumours. He points out that the doses current in radiosurgery (20 – 200 Gy) should reduce the probability of survival of irradiated cells to approximately 10^{-6}. He goes on to suggest this may produce adequate tumour control for all practical purposes. However, he adds a caveat that in vitro studies, from his and other laboratories have indicated a considerable variation in the radiosensitivity of both healthy and tumour cells, of the types likely to find themselves within a radiosurgical target volume

Conclusion

Current effort with the Gamma Knife is aimed at improving its range of application and therapeutic ratio. To this end, new dose strategies are being examined as are the use of various chemicals aimed at selectively increasing the radiosensitivity of the target organ. Moreover, the use of some degree of fractionation is also being considered in order to take advantage of the nervous system's high fractionation sensitivity and thus increase the doses which may be used on the pathological tissue. This is potentially most interesting in relation to larger lesions, like many arteriovenous malformations and meningiomas. Finally, it must be emphasised that the size limitation of the lesions to be treated in the Gamma Knife is a essentially a question of brain biology. Treatment of large lesions will be associated with a large integral dose to the surrounding brain, and the tolerated integral dose is a function of dose volume.

Suggested Further Reading

1. Arndt J, Backlund E-O, Larsson B, Leksell L, Norén G, Rosander K, Rähn T, Sarby B, Steiner L, Wennerstrand J (1979) Stereotactic irradiation of intracranial structures: physical and biological considerations. INSERM Symposium 12: 81–92
2. Backlund E-O (1979) Stereotactic radiosurgery in intracranial tumours and vascular malformations. In: Krayenbühl H et al (eds) Advances and technical standards in neurosurgery, vol 6. Springer, Wien New York, pp 1–37
3. Larsson B (1992) Radiobiological fundamentals in radiosurgery. In: Steiner L (ed) Radiosurgery baseline and trends. Raven Press, New York, pp 3–14
4. Larsson B, Lidén K, Sarby B (1074) Irradiation of small structures through the intact skull. Acta Radiol 3: 512–534
5. Leksell L (1971) Stereotaxis and radiosurgery. An operative system. Ch C Thomas, Springfield, I
6. Van der Kogel AJ (1990) Central nervous system radiation injury in small animal models. In: Gutin PH, Leibel SA, Sheline GE (eds) Radiation injury to the nervous system. Raven Press, New York, pp 1–26
7. Withers HR, Peters LJ (1980) Biological aspects of radiation therapy. In: Fletcher GH (ed) Textbook of radiotherapy. Lea & Febiger, Philadelphia, pp 103–180

II. The Patient's Experience

8. Gamma Knife Radiosurgery: A Patient's Eye View

Introduction

The patients requiring Gamma Knife radiosurgery will usually be suffering from a dangerous, potentially life threatening and thus intensely worrying condition. Yet where are they to receive information of the treatment, if their own doctor decides that this is the best option? There is a certain amount of highly specialised original literature, virtually no general knowledge and unfortunately a fair amount of rumour, more or less ill informed. Anyone who has treated even a small number of patients with radiosurgery will be able to confirm the role of rumour in the information these patients have received. Thus, it is necessary to have some hard reliable and readily comprehensible facts for the patients. Such information seems to come under two main headings. Firstly, there is information relating to the technique of treatment itself. This will be the subject of the current chapter. Secondly, there is information relating to the specific illness and the expected course and chances of success, following treatment. This seems more properly to be the province of the individual chapters related to the individual illnesses, which follow in the next section. Inevitably the suggestions for information that follow here are predicated on the assumption that the Gamma Knife has been chosen as the desired form of treatment.

Information at the First Consultation

The Nature of the Technique

Firstly, how does the Gamma Knife work, irrespective of the illness being treated? Well, to begin with it is not a laser and there is no requirement to open the head to apply it. An amazing number of people seem to regard a Gamma Knife as some sort of laser machine. Clearly, the mechanism of action has already been explained in some detail in previous chapters. However, not all patients are interested in this degree of detail. One analogy which is useful is to compare the Gamma Knife's focussing of gamma rays to the effect of a magnifying glass on the rays of the sun. An important feature to emphasise is that the amount of damage done with a magnifying glass, focussing the sun's rays depends on how long it is held in position and the same applies, in essence, to the Gamma Knife.

The Safety of the Technique

Patients are naturally concerned as to the nature of the rays involved. They are informed that gamma rays are like X-rays only more energetic. Like X-rays they penetrate the body, without it being necessary to open the body, so that surgery is avoided. However, such rays are clearly very powerful and thus potentially dangerous. The safety of the treatment is mainly related to the geometry of the machine which restricts damaging levels of radiation to such small volumes and enables precise determination of these volumes, avoiding normal brain. However, it is important to mention that a technique, capable of treating serious intracranial illnesses, is a powerful treatment and has unavoidably some complications. These are mentioned in more detail in relation to individual illnesses. Nonetheless, the most gratifying feature of the Gamma Knife to date is that while the treatment may have a low morbidity, it has a zero mortality.

The Degree of Disruption to the Patient's Life

In general, the patient is admitted on one day, treated the next and discharged on the third. Sickness benefit, related to the treatment procedure itself need not be extended beyond the weekend following discharge from hospital. Some would say even that was generous.

The Waiting

Both the patient and the referring physician will need to expect some degree of waiting, between the referral and the acceptance of the patient. This delay may be of the order of a few days to weeks and does not relate to any waiting list. The reasons for delay in response to referral are for the most part banal.

Firstly there is postal delay, especially if the patient is referred from abroad. It would seem for example to be a bad idea to send X-ray pictures rolled up in a convenient tube rather than in a large inconvenient envelope. The reason is that the tube could potentially contain weaponry and these tubes are kept back by customs officials, often for quite protracted periods. So avoiding X-rays in tubes will facilitate referral. A second and even more common cause for postal delay is incorrect name and address of the department to whom the patient is referred. This is obviously frustratingly avoidable for all parties involved.

The third delay relates to the X-rays which are sent. Ideally all fresh X-rays should be sent with the referral letter. The team assessing the suitability of patients for treatment have difficulty in making decisions of adequate quality on an arbitrary selection of pictures. For tumour patients MRI pictures are particularly valuable, in that they often show a greater degree of tumour spread than is visible on CT pictures. For arteriovenous malformation patients, high speed recent cerebral angiography is a must. It is particularly important to send films quickly if the patient is to come to radiosurgery following embolization, if maximal advantage is to be taken of the reduction in size produced by the embolization.

Another problem relates to language. The way services are organised at present it is not uncommon for Gamma Knife centres to have an international clientele. It is clearly desirable that the Gamma Knife units mobilize as many languages as possible around their machine. However, it is probably better to refer a patient in bad English rather than fluent Greek or Spanish, until the referring physician knows for sure that the receiving specialist will be able to understand the referring letter. Attempts will clearly be made to reply in the language of the country of origin of the referred patient, wherever possible. But English remains the *lingua franca,* when in doubt.

Departments engaged in radiosurgery are few, though rapidly increasing in number and the colleagues engaged in this work are all more or less acquainted with each other. The author has no difficulty in being able to affirm that there is a great commitment to try to get the patient to the treatment as quickly as possible. The factors mentioned above are a few ways in which referring physicians can avoid unnecessary delays.

Finally, in respect of delays, there remains the waiting list for patients whom have been accepted for treatment. For this there is no simple answer beyond increasing the number of centres using Gamma Knives. Certainly, the experience from the centres that have treated patients on a regular basis for some years is that the waiting list tends to increase in length.

Information on Admission to Hospital

The Brochure

In Bergen, all patients accepted for treatment are sent a brochure covering the major steps of the procedure, at the same time as they receive a letter confirming their appointment. At the present time this brochure is printed in Norwegian, English and German and an Italian edition is under preparation. This is mentioned to confirm the importance of communicating with the patient in his/her own language in what will often be a multilingual undertaking.

The Day Before

The patient is admitted to the department in the usual way and a clinical examination is undertaken. Then a detailed explanation is given which has two main elements. Firstly, the aims of treatment, the risks and the expected final results are discussed. Thereafter, the sequence of events that the patient will undergo the following day is explained. It goes without saying that children are spared much of this information, since the procedure in children is undertaken under general anaesthesia. It is usual to explain that most of the treatment is painless. However, the stereotactic frame application is uncomfortable or even painful for a very few minutes, either at the site of the pins which fix the frame to the head, or in the external auditory meatus where steadying pins are applied to secure accurate placement of the frame. Though the author has not used these pins for over two years.

The Day of Treatment

The patient may receive premedication, though this should be done with some judgement and not as a routine. If necessary 10 mg diazepam is useful as a sedative. On the other hand 4 mg dexamethasone, half an hour prior to applying the frame is useful in relieving the nausea and vomiting that some patients experience after radiation treatment. In some cases, with critically placed large tumours, the surgeon may wish to place the patient on full doseage dexamethasone, for a couple of days, followed by gradual withdrawal of the drug over a week or so. The purpose of this is to reduce the risk of acute swelling following Gamma Knife surgery. It should be stated that at the present time this is merely a precaution. To the best of the author's knowledge it is extremely rare for such a swelling to occur with this technique.

The frame is applied with the patient sitting in a chair or bed. The hair is washed in an alcohol based disinfectant solution. No shaving of hair is required; a popular detail this. After the frame is applied the rest of the day is essentially tedious. If an arteriovenous malformation is to be treated an angiogram follows and this may be slightly uncomfortable, though not markedly so with modern contrast media. Otherwise, a CT or MRI examination is performed, following which the patient is returned to the ward, while the Gamma Knife team carries out dose-planning on the basis of the X-rays taken. When the dose-planning is completed the patient comes to the Gamma Knife suite and treatment is performed. This may involve several turns in and out of the machine, depending on the site and shape of the lesion. The actual treatment while again rather tedious is entirely painless. After it is over the frame is removed, and a bandage applied to the head. If the patient has undergone angiography this will determine the period of immobilization. Other patients may mobilize at their convenience. A few will suffer an acute short term pain reaction following removal of the frame This is best managed with reapplication of the head bandage with less compression and if necessary one intramuscular injection of an adequate dose of an opiate analgesic. Further post-operative treatment than this should be unnecessary.

The Day after Treatment

This is the best time to give information on the further outlook and especially the plans for a follow-up, which is necessarily detailed and long term. This is also the time to discuss unwelcome restrictions in activity, though clearly such decisions are more appropriately the responsibility of the referring department. Finally, a very few patients may experience pain related to the fixation of the frame, especially where one of the posterior pins has been in contact with an occipital cutaneous nerve. Explanation of the reason for this discomfort and that it is due to the frame and not to some unwelcome intracranial complication seems to go a long way to reconciling the patient to the discomfort which would appear to be almost invariably short term.

III. Clinical Aspects

9. Diverse Clinical Aspects

Introduction

This chapter is concerned with sundry factors related to patient assessment and Gamma Knife surgery, but unrelated to a specific diagnosis. There is a widely accepted assumption that the radiosensitivity of cerebral tissue, in different locations within the brain varies. The truth of this assumption and how it should affect dose-planning is clearly important. Moreover, a number of the lesions treated in the Gamma Knife are located at the base of the skull, close to a variety of important extracerebral structures. The tolerances of these structures are also a source of concern for the user of the Gamma Knife. Then, it is important to consider the influence of the volume of the lesion on the results of treatment on the lesion itself and on the tissue around. Another area which may give rise to confusion is the effect of the Gamma Knife on symptoms. It is a commonplace that a patient seeks a doctor to obtain relief of a symptom or symptoms but is treated for a disease: while the symptoms may or may not be alleviated. This discrepancy between the patient's wishes and the patient's best interest is also relevant for those who require Gamma Knife surgery.

Variations in Intracerebral Radiosensitivity

The early work on the cerebral doses required for the reliable production of an intracerebral lesion is recounted in chapter 6. It was found that a dose of 150 to 200 Gy was needed to produce an **acute** localized, non-progressive cerebral necrosis which could be used for clinical lesion production. Moreover, in experiments performed prior to the clinical application of radiosurgery, it was shown that the chance of obtaining *acute* necrosis with a dose below 150 Gy were negligible. On the other hand **delayed** damage may be produced with a single dose in excess of 20 Gy, but is unlikely below this dose level. It is this delayed damage, appearing clinically at about six to twelve months after treatment, that concerns the surgeon.

These dose levels give some guidance for the Gamma Knife surgeon. However, all these doses apply to normal brain. Much of the work responsible for the data described in the previous paragraph has been carried out in normal nervous tissue in experimental animals. The Gamma Knife surgeon has to ask the question is the pathological brain around a lesion as sensitive or rather as insensitive to irradiation as normal brain? Moreover, the degree to which cerebral function is concentrated or dispersed varies greatly through the brain. One of the reasons that patients are referred for Gamma Knife treatment is to avoid surgical trauma to regions, where there is a great concentration of vital function. This is particularly true of the brain stem, not least because this is a region which controls vegetative functions. There is a general tendency to reduce the integral dose in these very sensitive regions. The requirement for this caution was demonstrated by Steiner at a symposium in Charlottesville in 1989, where he demonstrated that the therapeutic index for the treatment of arteriovenous malformations was quite small. Work is at present under way in a variety of centres to define and quantify the tolerable brain integrated dose for different locations and different lesions. However, this aspect of treatment will continue to be characterized by a degree of uncertainty for some time to come.

Variations in Extracerebral Radiosensitivity

A significant proportion of lesions treated by the Gamma Knife either lies in close relation to the cerebello-pontine angle or to the sella turcica. In both these regions there is a risk of cranial nerves receiving a substantial dose of radiation In this context it must be emphasised that two of the relevant nerves are not normal nerves at all. The optic nerve and chiasm are, as is well known, a projection of the brain. However, even if they are not truly extracerebral tissue, they are convenient to consider in this section, because of their location. The auditory nerve has a special anatomy and amongst its special features is an absence of neurilemmal cells. The importance of vision and the known vulnerability of the acoustic nerve to mechanical manipulation has led to these two nerves being treated with great respect. While it has not been possible to prevent damage to hearing in many cases, to the best of the author's knowledge, treatment of pituitary tumours and craniopharyngiomas has not led to loss of vision, as a result of Gamma Knife irradiation. Visual deterioration has been noted in one or two cases, though it is still too early to know if this is a permanent change or only temporary. The permissible dose to these two nerves remains a matter for debate. However, the clinical evidence indicates that 10 Gy or less is usually well tolerated.

In contrast with these two special nerves the nerves related to the cavernous sinus region, particularly those concerned with external ocular movements, seem to be remarkably resistant to Gamma Knife surgery. This is in keeping with their ability to function following mechanical manipulation, during open surgery. Deficit arising from such manipulation is commonly reversible. On the other hand sensory deficit, related to trigeminal irradiation, may be somewhat more intransigent. This is in keeping with the generally applicable neurosurgical finding, that sensory neurological deficit, due to involvement of a peripheral nerve or nerve root, is more persistent than is a motor deficit.

Other structures that could give rise for concern at the base of the skull are of course the major blood vessels. There is for example no way a cavernous sinus meningioma can be given an adequate radiation dose without the carotid artery receiving a major dose of radiation as well. However, there are two pieces of evidence suggesting that this may not be as important as at might appear at first sight. Firstly, there is work done on cats, with very high radiation doses delivered to the normal basilar artery, with a Gamma Knife. The doses varied from 100 Gy to 300 Gy. While marked radiation lesions were seen in the walls of these normal arteries it seemed virtually impossible to obtain arterial occlusion, at least within the time frame of the experimental design. The second piece of evidence that suggests that normal arteries tolerate relatively high doses is the finding that occlusion of arteriovenous malformations does not seem to be accompanied by occlusion of neighbouring normal arteries, at least not in most cases. These two sets of observations suggest that normal arteries are relatively radioresistant to the doses that are relevant for radiosurgery.

The Effects of Lesion Volume on Dose Planning

As has been pointed out in chapters 4, 5 and 7, the larger the treatment volume the greater the chance of damage to normal tissue around the lesion. Moreover, it is the conventional wisdom that the larger a malignant tumour to be treated with radiotherapy, the smaller the chance of success. This does not of course mean that larger tumours are necessarily less radiosensitive, though they may be. However, larger tumours place more severe constraints on the dose-planner, irrespective of the radiation technique involved. Furthermore, since the relation between dose and cell survival is a matter of chance, it follows that the larger the number of cells being treated, with a given dose, the smaller the chance of obtaining a proportion of cell deaths adequate to enable tumour control. Thus, larger lesions are always liable to be less satisfactory radiosurgical targets than smaller lesions. Moreover, the geometry of the Gamma Knife is such that it is more difficult to obtain a sharp dose fall to match the edge of a lesion with larger lesions. Even so, the biggest problem with larger lesions is the increasing radioresponsiveness of large volumes of brain tissue, as compared with small volumes (see chapter 7). The relationship between lesion volume and successful treatment will be taken up again in the chapter on arteriovenous malformations.

The Effect of a Peripheral Intracranial Location on Treatment Planning

The design of the Gamma Knife is such that it is much simpler to treat lesions that are centrally placed in the brain. While the Leksell frame's axes run from values of 30 to 170 the Gamma Knife's axes run, with some slight variation, from only 50 to 150. In other words, the Leksell frame's axes extend for 140 mm in each of the 3 directions, while the Gamma Knife's extend only in 100 mm. Figures 9.1–9.4 illustrate some strategies whereby this problem may be overcome. Firstly, the frame is applied eccentrically, in an attempt to place the lesion as near the centre of the axis system as possible. Secondly, the frame may be placed low down on the head to increase the access to eccentric lesions. Thirdly, the frame may be rotated on the head. These techniques have proved sufficiently successful that it has been possible to treat lesions at the foramen magnum, in the retina of the eye and close to the frontal and occipital poles. Nonetheless, it should be born in mind, that the treatment of peripheral lesions, in particular when using the high doses required for metastases, that there may be some risk of skin damage. To date, concern for complications related to cortical damage with peripherally placed targets does not appear to be justified. The only changes of which the author is aware, at this time, is a temporary focal alopecia. Finally, it should be noted that the strategies outlined in this section, to enable access to peripherally placed lesions will not work for all patients. While it is extremely uncommon, even so, the banal factor of the size of the patient's head may assume critical significance. Thus, it may be essential to request X-ray pictures which enable measurement of the distance from the centre of the proposed target to the skull. The author, on one occasion, has had to abandon treatment in a patient with an optic melanoma and a very large head.

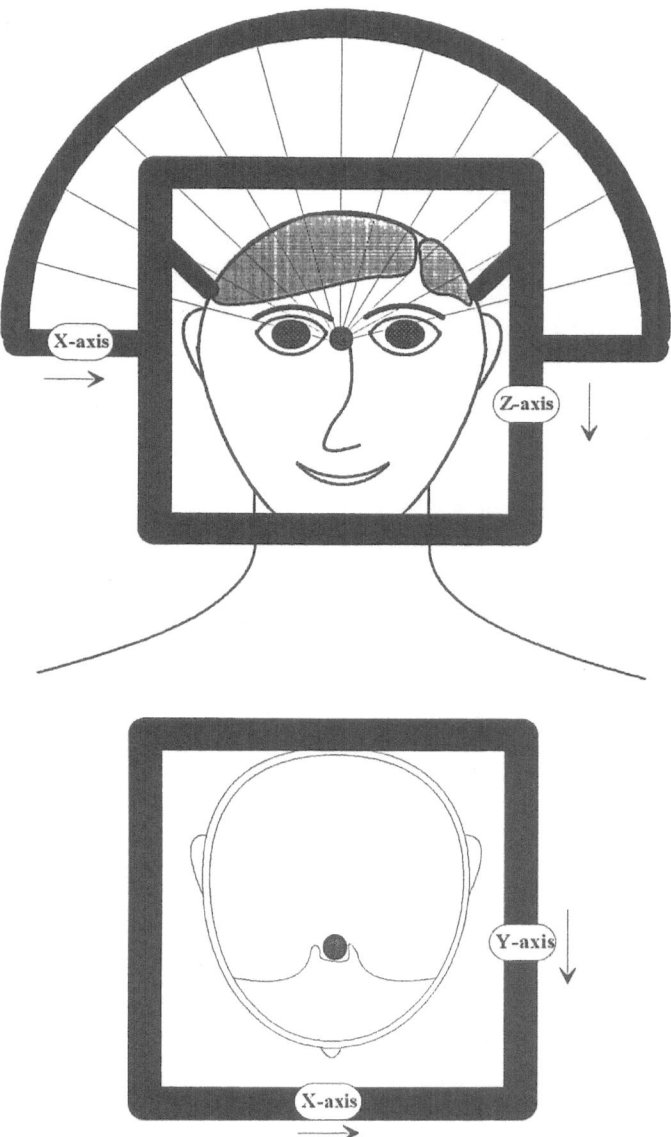

Fig. 9.1 Pituitary Adenoma
Diagram concerning frame placement for a pituitary adenoma. The upper picture represents the head as fixed to the Gamma Knife helmet. The thickly drawn square represents the frame. The oblique placed bars inside the frame represent the pins fixing the frame to the head. N.B. they are not fixed in the same plane as the target. The lower diagram represents the placement of the skull and target with the contained section of the head, as seen on a CT pitcure. In this sort of case the head is placed far back in the frame and centred

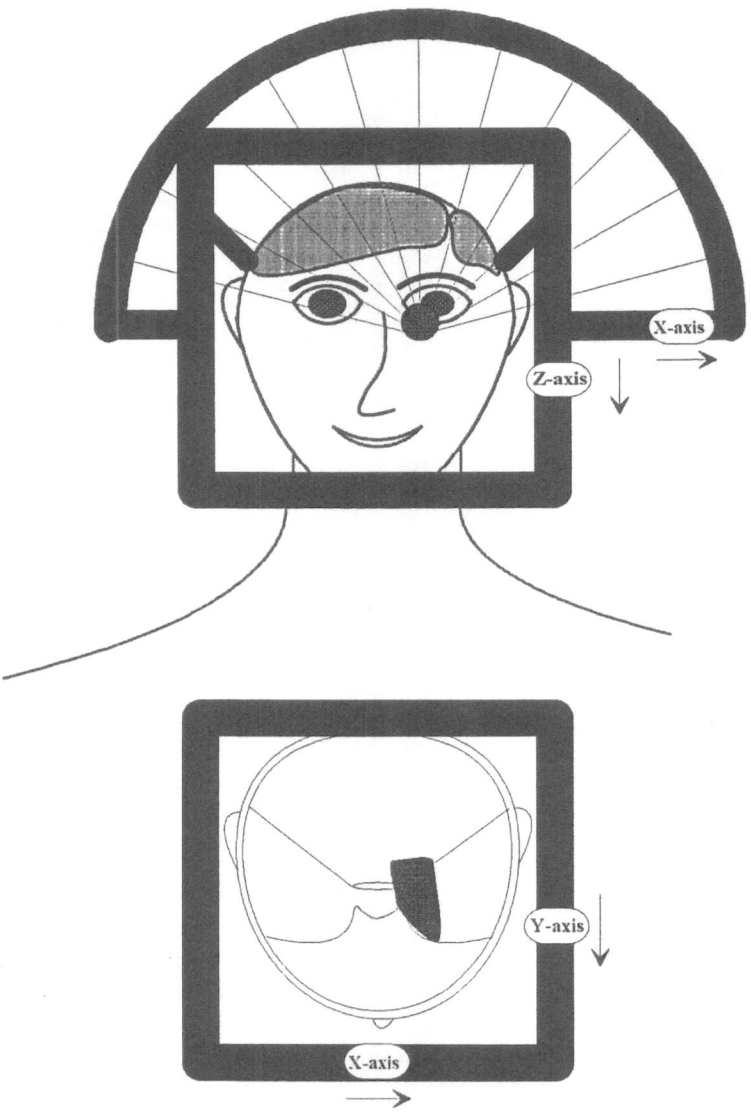

Fig. 9.2 Cavernous Sinus Meningioma
Diagram concerning frame placement for a cavernous sinus meningioma. In this picture the head in the lower picture can, with this anterior lesion, be seen to be displaced backwards in the frame. It is in both pictures displaced slightly away from the side of the lesion. Thus, while the lesion is centred to some extent around the lesion this is not maximal. This only slight attempt at obtaining a centred lesion, together with the relatively high fixation of the frame on the head leaves the frame in contact with the Gamma Knife helmet on the right side. Thus further lateral adjustment in that direction would be impossible in this case

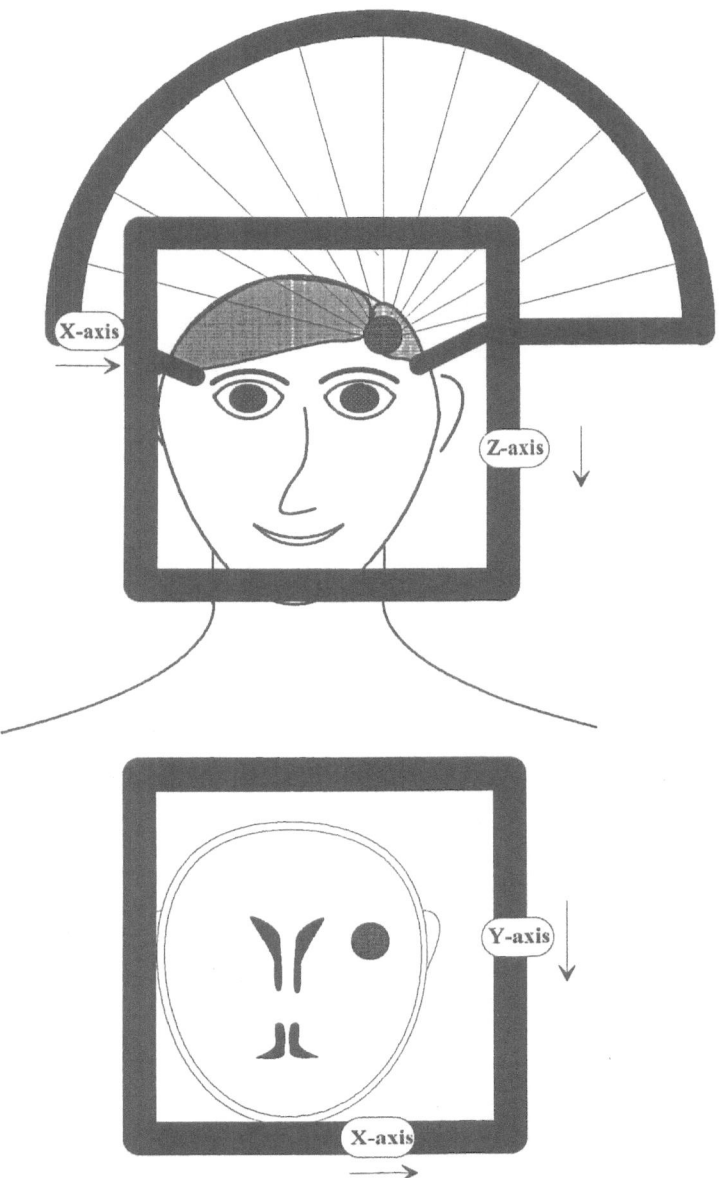

Fig. 9.3 Arteriovenous Malformation

Diagram concerning frame placement for an arteriovenous malformation. In this case the AVM is a very laterally and rather posteriorly placed. Thus the head is displaced laterally towards the opposite side from the lesion as in Fig. 9.2. Moreover it is displaced forwards, this time. In addition the low fixation of the frame on the head means that the head is projecting further out of the helmet than in Fig. 9.2. In consequence, the frame is not in contact with the helmet, and there is still some room left over for further lateral adjustment if needed

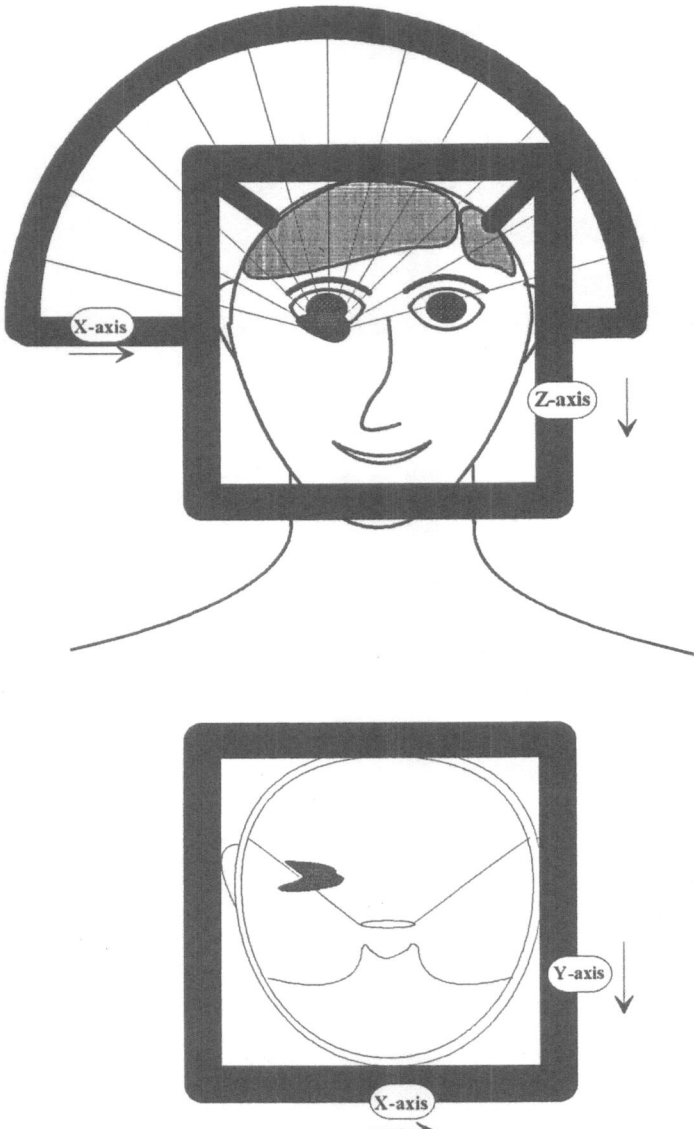

Fig. 9.4 Acoustic Neurinoma

Diagram concerning frame placement for an acoustic schwannoma. In this case the head is shown to be too large for much manoeuvring in the frame to be possible. However, with a low frame fixation it is still possible to place the lesion adequately centrally and obtain a successful treatment. Note that, in this as in the other diagrams in this chapter, the Gamma Knife radiation is directed at a particular focus all the time. The purpose of the dose planner is to manipulate the head so that the given dose is placed concordant with the focus of the machine

The Effects of Treatment on Symptoms

It is, as stated above, a common doctor's dilemma that the patient requires relief of symptoms while the doctor offers treatment of a potentially dangerous disease. The doctor has little choice in his course of action but it is incumbent on him to explain to the patient why the treatment has been a success, while the patient feels little better. This is also true for the Gamma Knife, in that its effect on symptoms may be, in some cases, disappointing and successful treatment is mostly recorded in terms of X-ray changes. However, there are exceptions. Occasionally, hearing may improve with an acoustic neurinoma patient, but this is the exception. In over 50 % of Cushing's disease, persistent Gamma Knife treatment will result in symptomatic relief. There is also a tendency for improvements in epilepsy especially for patients with arteriovenous malformations. In most cases, the patient will notice little difference. For most arterio-venous malformations the benefits are not so much improvement of symptoms as removal of the threat of rebleeding. There is some confusion in this respect amongst both physicians and patients. Thus, it is the experience of the Bergen group that pain, related to a tumour or secondary to the damage produced following haemorrhage from a malformation is not relieved by treatment, even if the radiological result is satisfactory. Moreover, it would appear to be as difficult to reduce or abolish tinnitus with the Gamma Knife as it is with other modes of treatment. It is clearly important that the patient is made aware of the aim of treatment which is primarily to control the lesion not to relieve the symptoms. However, if symptom relief does occur this is of course a welcome benefit.

It might be thought that the inability to reduce symptoms might discourage patients from seeking Gamma Knife surgery. This is not the experience of most users, for a number of reasons. Firstly, other techniques, for example microsurgery, may also be unpredictable in their effect on, for example tinnitus, due to an acoustic neurinoma, or epilepsy due to a meningioma. Secondly, the gamma knife is often used to prevent the development of new, treatment-related symptoms. This is particularly true in the treatment of deep-seated central arteriovenous malformations or cerebral metastases. Thirdly, the very thought of an intracranial lesion is so distressing that people often display a considerable capacity for tolerating persisting symptoms, so long as they can be assured that their lesion no longer represents a threat. However, not all patients react this way. Thus, it remains important that the Gamma Knife remains in the hands of competent neurosurgeons. In a proportion of patients there will be alternative surgical options and these should be available. This is one of the reasons that the same person should be proficient in both conventional surgical and radiosurgical techniques. There is a second reason for such double proficiency. The Gamma Knife technique is based on the assumption that the radiation dose is distributed in such a way as to deliver a maximum to the pathological target, while sparing normal tissues as far as possible. The ability to accomplish this is in turn predicated on the assumption that the responsible doctor has a thorough knowledge of the three dimensional anatomical relationships of the region concerned. This knowledge and experience is only gained after several years in the operating theatre. It is a commonplace, in the training of young neurosurgeons, that it is difficult to orientate oneself at surgery within the head. This is a skill which cannot be learned from a book or from the perusal of X-rays. However, it is vital in planning radiation doses, in relationship to normal anatomical structures which are not clearly delineated, even with the most sophisticated of modern imaging techniques. Thus, if the capabilities of the Gamma Knife are to be exploited to the full, it should be in the hands of neurosurgeons with adequate experience in the surgery of the regions which they are treating with the Gamma Knife. If this argument is respected, the patient may be offered an optimal treatment, whether it be microsurgical or radiosurgical; a treatment tailored to individual requirements and psychological needs. This principle of treatment is very much in keeping with Leksell's original teachings and is a part of the tradition of Gamma Knife surgery.

10. Cerebral Arteriovenous Malformations

Introduction

Cerebral arteriovenous malformations have been most gratifying lesions to manage with Gamma Knife surgery. The treatment was started by Professor Ladislau Steiner, at the Karolinska Institute in Stockholm, in 1970 and since that time he has treated over 600 patients with very good results. However, the variability of the lesions and the size limitations imposed by brain tolerance to radiation make it impossible to offer Gamma Knife surgery to all patients with an arteriovenous malformation. Indeed, Steiner emphasises, in the Leksell tradition, that the Gamma Knife user must also be adequately expert in the surgical removal of the lesions he is treating. Moreover, the introduction of improvements in angiographic catheters, with corresponding improvements in embolization techniques means that a variety of treatments are available: thus, the treatment chosen should be tailored to the patient's needs. Moreover, the different treatment methods may be combined to provide optimal management. In consequence, cooperation between departments and across speciality boundaries is an important aspect of the management of this condition.

Nomenclature

Cerebral arteriovenous malformations comprise a group of blood vessel malformations. There is general agreement, at the present time that capillary and venous malformations are not suitable objects for radiosurgery. The arteriovenous malformation is considered to be a congenital anomaly, with shunting of blood directly from arteries to veins, without intervening capillaries. Thus, these lesions are also called fistulas. The shunt is often localised to a nidus; consisting of a network of irregular, pathological blood vessels, with cerebral parenchyma in the interstices. Their total removal should result in restitution of the surrounding circulation to normal (Figs. 10.1 and 10.2). Because there is a localised, apparently space occupying lesion, these malformations were also called angiomas in earlier days. How-

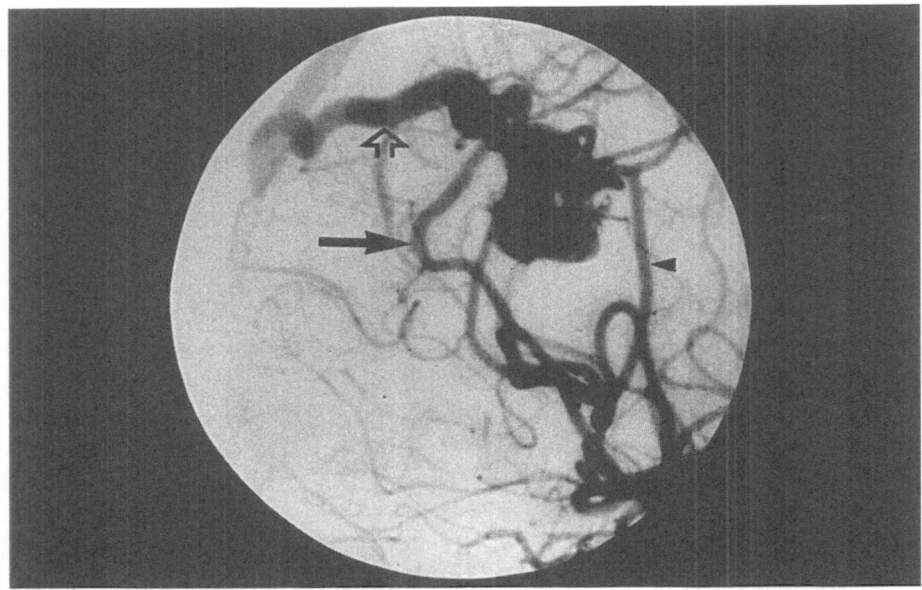

Fig. 10.1. a

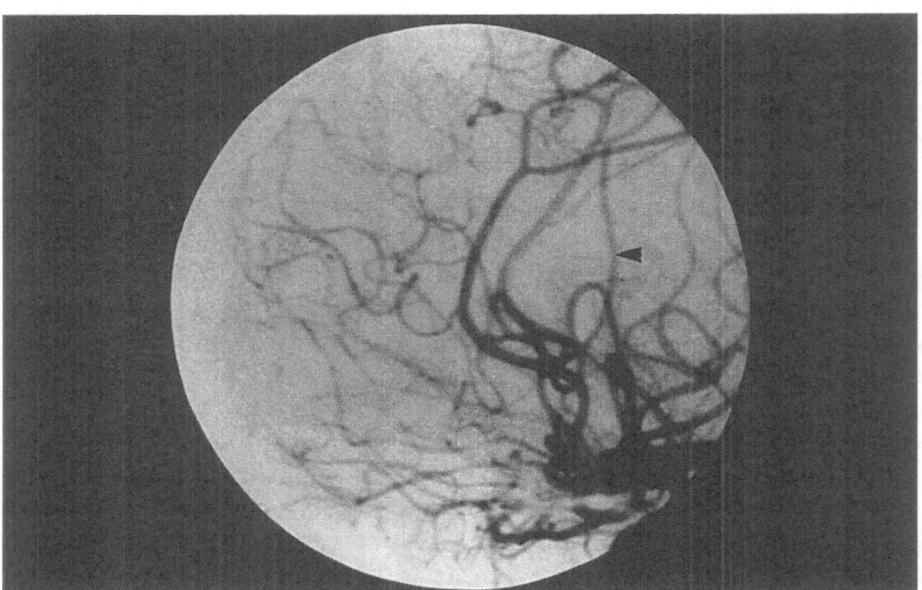

Fig. 10.1. b

ever, as stated above, there is a consensus today that they are not neoplastic but congenital in origin. The arteriovenous malformations will subsequently be referred to by the abbreviation AVM.

Arteriovenous Malformation

Clinical Presentation

The majority of AVMs present clinically with haemorrhage. The next most common presentation is epilepsy. A relatively small minority present with headache, ischaemic symptoms or as a chance finding. Haemorrhage may be subarachnoid or intracerebral or both. Factors that predispose to rebleeding have been much debated over the years. There is agreement that one bleed is the single factor most likely to predispose for a second. Moreover, pregnancy and advanced age at presentation also increase the likelihood of rebleeding. It may also be true that smaller malformations bleed more often than large but this finding is reported variously by different authors.

Aims of Treatment

Since the effects of any form of treatment is uncertain in respect of the symptoms associated with the condition, the main reason for treatment is prophylactic, to prevent the dangers of repeated bleeding. Thus, since the therapy is essentially preventative, it must be a sine qua non that its consequences are not worse than those of the untreated disease. It is a characteristic of AVM bleeding that the mortality of an individual bleed is relatively low – about 6% – and that the rebleed rate

---

Fig. 10.1
(**a**) Pre-treatment angiogram of a young female patient with a right frontal AVM. This patient had never bled. She was offered the choice of surgery or the Gamma Knife and chose the Gamma Knife largely due to cosmetic considerations. The malformation is mainly supplied by two middle cerebral artery branches – an anterior (arrow) and a posterior (arrowhead). Note that these feeding arteries have a larger diameter than other equivalent arteries, indicating hyperdynamic flow. The AVM drains into the superior sagittal sinus through a large early filling vein (open arrow). (**b**) New angiogram, two years after Gamma Knife treatment. Note that there is no AVM, no early filling vein. Moreover the anterior feeding artery has disappeared while the posterior feeding artery (arrowhead) now has a diameter commensurate with other equivalent arteries

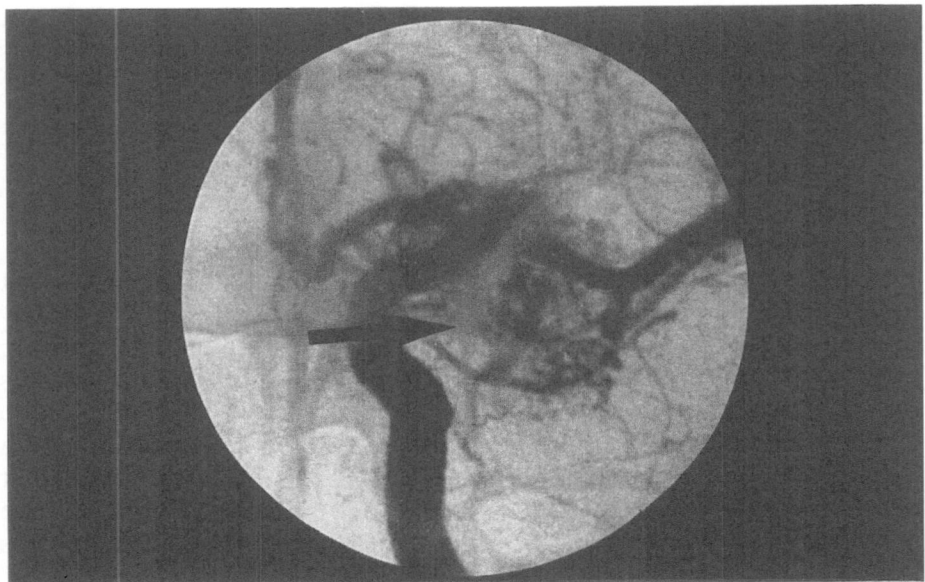

Fig. 10.2. a

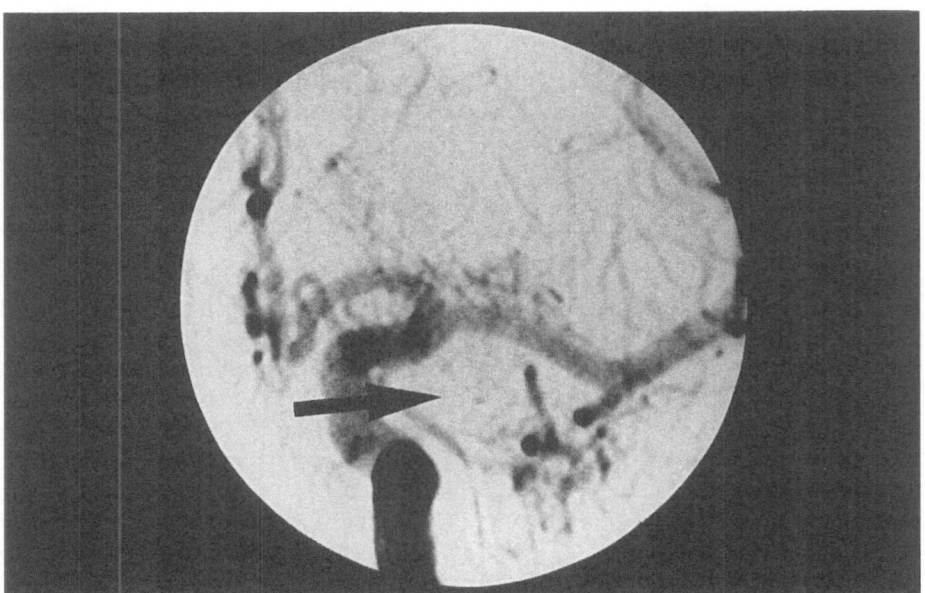

Fig. 10.2. b

is about 3% per year. Thus, the high mortality associated with hypertensive intracerebral haemorrhage and aneurysmal subarachnoidal haemorrhage is not a feature of AVMs. On the other hand they occur in a lower age group than the other major causes of intracranial haemorrhage so that there is a longer life expectancy for the accumulation of rebleed risk.

Since the risks of treatment should not exceed the risks of the natural course of the disease, it is important to have a clear idea of this natural course. Unfortunately, this is very difficult, because to randomize a treatment plan for treatable AVMs would be considered unethical. This attitude is based on by no means randomised retrospective studies. These report a variety of results but, if attention is restricted to the larger studies, then a rebleed rate of about **3 % per year** for AVMs which have bled once or more, and of **about 1 % per year** for those which have not bled, seems to be a reasonably constant finding. The mortality varies from series to series but seems to be more than 20 % over a twenty year period for both patients with bleeds and for those with epilepsy.

Thus, the aim of treatment in the Gamma Knife is to achieve total AVM obliteration with a substantially better morbidity and mortality than that quoted in the previous paragraphs. It must be emphasised that it is the belief of all major centres involved in Gamma Knife surgery, that nothing less than **total obliteration** will reduce the rebleed rate. It is necessary to emphasise this because the Boston group, who use proton irradiation in the treatment of AVMs suggest that reduction in AVM volume can contribute to reducing the risk of rebleeding. This matter is not resolved at the present time and it seems correct to be cautious and insist on total obliteration, before announcing that a cure has been achieved. It must also be mentioned that if a nidus has disappeared but an *early filling vein* (Fig. 10.3) is still visible, then total obliteration has not been achieved in the view of most workers.

Fig. 10.2
(a) Angiogram from a patient with a partially resected left temporal AVM (arrow). Partial resection, prior to Gamma Knife surgery is becoming increasingly used. With modern operation and diathermy techniques it would appear to be safe. It would also seem to be more effective than embolization as a pre-treatment for the Gamma Knife because it seems more certain to reduce the AVM's volume, while the embolization often only reduces the flow. (b) Angiogram one year after treatment from the same patient. This young man was a marathon runner and unwilling to wait two years before his control angiogram. The malformation has completely disappeared. He is successfully competing in marathon events again

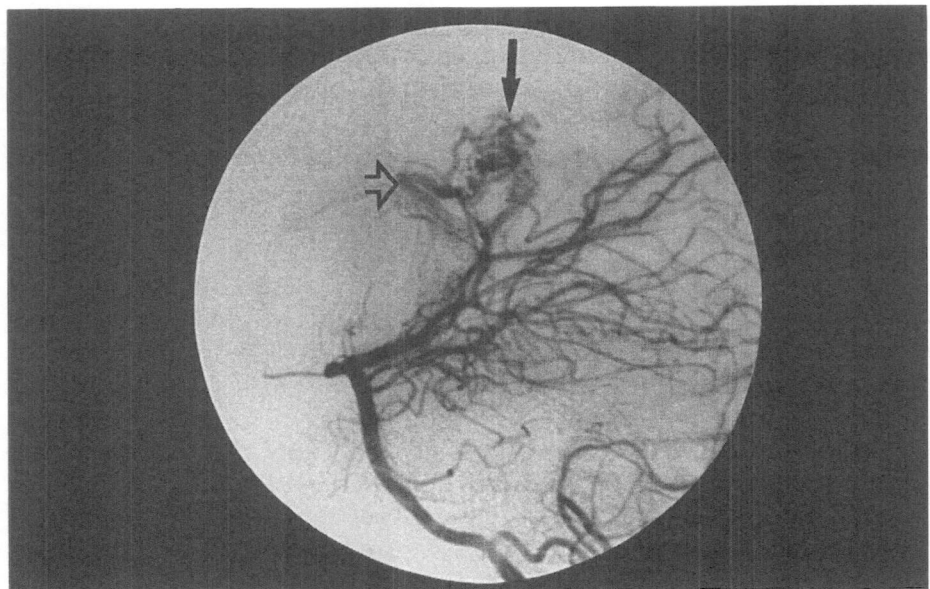

Fig. 10.3. a

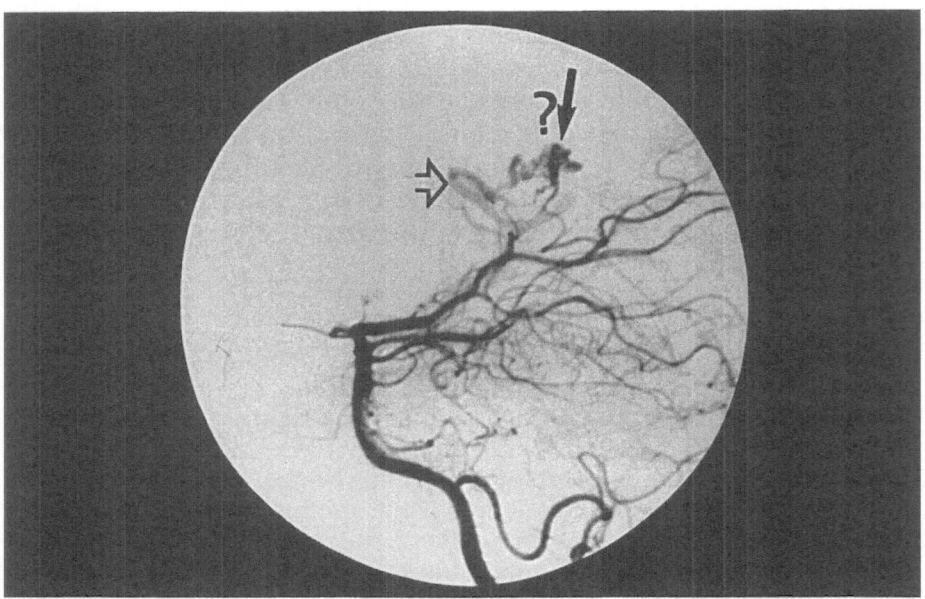

Fig. 10.3. b

Changes in Arteries Following Radiation

As has been mentioned earlier in this book radiation has an effect on any biological tissue, normal or pathological. However, different tissues have different radiosensitivity. It would appear that arteries are only moderately radiosensitive. Many studies, both clinical and experimental, have been undertaken to examine the effects of radiation on arteries but in many the tissues and their arteries receive the same dose. In clinical studies of this nature, analysis of the interaction of the radiation, local infections and the effects of the neoplasm undergoing treatment is very difficult. In experimental studies where there is no dose discrimination between the irradiated tissue and its supplying blood vessel it is not easy to assess how much of the observed effects on an arterial wall are the result of primary radiation damage and how much are secondary to changes in the parenchyma. It is also necessary to appreciate that the blood vessels under consideration here are arteries, not the small arterioles and capillaries involved in vascular radiation damage to the tissues.

To date, two experimental studies have used stereotactic localization techniques to aim a high focal radiation dose at a particular artery, to study the effects of radiation on normal arteries. In the first of these studies, by Nilsson et al., in cats, the doses were so high (100 to 300 Gy) as to be totally inappropriate for clinical practice. They were in fact so high as to produce lesions within a matter of days to weeks. These are considered to be high dose, short latency delayed reactions; the short latency being due to the very high dose. The arterial changes consisted of endothelial damage, degeneration and necrosis, intimal swelling, elastic lamina fragmentation, media muscle cell swelling vacuolation and necrosis, and adventitial fibrosis. The arterial lesions were observed at varying times after the radiation

---

Fig. 10.3
(a) An AVM in the splenium, prior to treatment treated in the Gamma Knife. This awkward location for surgical attack is a popular location for Gamma Knife surgery. The malformation is indicated by the arrow and the early draining vein by the open arrow. (b) Angiogram from the same patient, two years following treatment. The malformation is much smaller and quite difficult to be sure if there is any residual nidus (question mark and arrow). However the virtually unchanged early filling vein (open arrow) indicates that there must be a residual nidus. In this case it would be appropriate to wait one more year and repeat the angiogram. If the vein is still present then a re-treatment would have to be considered

was delivered. There was some tendency to the development of hyaline degeneration with time. *Arterial occlusion was not seen.* There was however an early development of cerebral necrosis, after a few weeks, occurring in neighbouring parts of the brain which received a high radiation dose. These parenchymal changes may be considered similar to those observed at radiosurgical thalamotomy. In the second study, Kihlström, Lindquist et al. used doses from 10 to 100 Gy in rabbits. No vascular pathological changes, arterial, capillary or venous were observed with follow up of up to 17 months.

The lesions seen in clinical practice are different. They follow a dose of 20 to 50 Gy to a collection of pathological blood vessels in an AVM. Occlusion of the AVM occurs after about 1 to 2 years. It is not clear whether these lesions to AVM blood vessels are of the same kind as the short latency lesions described above, being merely delayed in time and reduced in intensity. Another view would be that they are qualitatively different. This is in accordance with the consistent observation that, in keeping with the high dose study on the cat basilar artery, normal arteries in the neighbourhood do not seem to occlude, while those of the AVM itself do. The mechanisms underlying AVM blood vessel occlusion are, as yet, not fully explained.

The only currently available information on the effects of therapeutic doses of radiation on AVMs comes from pathological specimens. These may be available following death from haemorrhage or from intercurrent surgery, following a rebleed after radiosurgical intervention. At the time of writing such a study is not available for the Gamma Knife but a relevant experience has been published following Bragg Peak therapy in Boston. Here, five brains which contained AVMs, which had been treated by proton beam therapy were examined. In two of these, examination occurred late enough for radiation effects to be visible. They consisted of narrowing or occlusion of malformation blood vessels by collagen. In a third patient examined following a fatal haemorrhage, six months after treatment, it was impossible to distinguish between radiation and haemorrhagic damage. However, tissue within 3 mm of the radiation field appeared to be normal. Thus, clinical doses with the proton beam technique can produce occlusion of malformation blood vessels, associated with the formation of collagen.

A number of central questions remain unanswered. In particular, further histological studies are needed to elucidate the nature of the occlusion process, occurring in AVMs after radiosurgery. Moreover, it remains unclear as to whether normal blood vessels are less radiosensitive than malformation blood vessels. The clinical evidence would suggest that they are because to date, to the best of the author's knowledge, no evidence has accrued to show that occlusion occurs in normal major cerebral arteries, following the therapeutic radiation doses currently in use in Gamma Knife radiosurgery. It also remains unclear as to what extent haemodynamic factors affect occlusion. Nor is it clear as to what extent such factors affect the reaction of surrounding cerebral parenchyma to radiation.

The Role of Radiation Treatment

Even though the mechanism of action of radiation on AVMs remains unclear, even today, the use of radiation for the treatment of these lesions has a respectably long history. The first patient to be treated with radiation, with reduction in the frequency of pre-treatment epilepsy, was reported in 1921 but irradiated in 1914. That radiation had an effect was well documented by the 1960s but, unfortunately, its effect was too slight in many cases and too uncertain for conventional radiotherapy to play any part in the management of cerebral AVMs.

Steiner, who as stated above pioneered Gamma Knife surgery of these lesions starting in 1970 and who now has treated many hundred patients, has shown that the treatment has a very definite place in the therapeutic armamentarium. The results to be reported here relate to total covering of the AVM by the radiation field. Other strategies have been attempted but they are less successful.

Results of Gamma Knife Surgery

In Steiner's hands the obliteration rate at one year of approximately 50%, with a long term success rate of 87% follows optimal treatment. The delay in occlusion is associated with a rebleed rate of approximately 3% per year prior to obliteration. In addition, there is approximately a 10% chance of a post-treatment, radiation-induced increase in symptoms. However, this deterioration is permanent in only about 3% of patients. This means that for the majority of patients there is about a 90% chance of being successfully treated with a management morbidity of about 10%. This is considerably better than can be expected from the untreated disease and thus the results meet the criteria of the aims of treatment as defined above.

Steiner has indicated that the therapeutic index between successful obliteration and radionecrosis is small. The quantification of the risk associated with treating an AVM of a given size is the subject of much interest. The Pittsburgh group have developed a formula designed to assess if the risk of radiation damage exceeds 3% or not. Moreover, the group is working on publishing more information

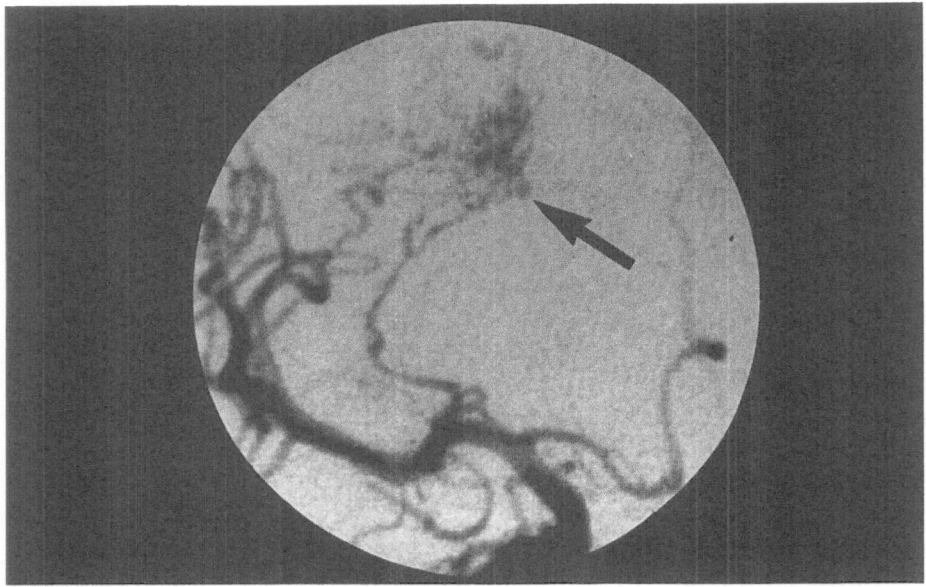

Fig. 10.4. a

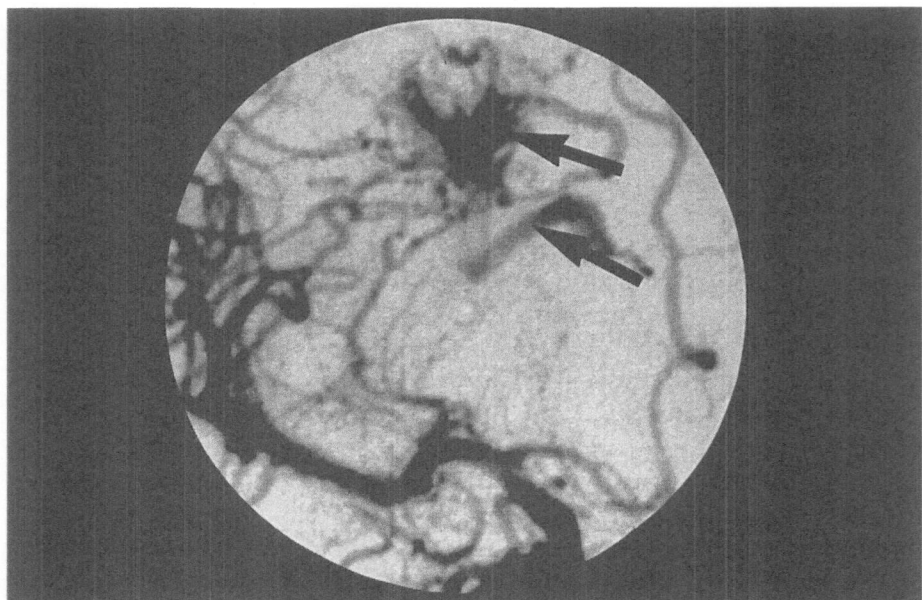

Fig. 10.4. b

related to this subject. It has also been indicated that smaller malformations have a better result than larger and that this does not appear to be dose related. This finding also applies for linear accelerator radiosurgery and proton beam therapy, so it would appear to be a basic radiobiological phenomenon, rather than a technical limitation of the Gamma Knife.

Patient Assessment for Gamma Knife Surgery

The Gamma Knife surgeon will require relevant computer imaging studies from the time of the presenting bleed. MRI studies are particularly useful for defining the areas of brain which have been damaged, following bleeding. This can be most useful information when designing a dose-plan. However, high resolution, high speed cerebral angiography remains the examination of choice in this condition. It is the only examination which gives resolution of the lesion in time. This is very important, because while the draining veins may be the largest structures associated with a malformation, the decision to treat or not is based on the size of the nidus, which is often only clearly defined on the earliest pictures (Fig. 10.4). It is relatively common to find that the actual size of an AVM, at the time of Gamma Knife dose-planning is a little smaller than was assessed on the pre-treatment films. This is because high speed angiography is consistently used in this context and a complete angiogram series is available and not a selection of pictures. In other words, it is very helpful if the referring physician sends the whole series. In general, any malformation that is not more than 25 mm in all its diameters will be accepted, though the chance of success would appear to decrease with the larger lesions. Nonetheless, larger lesions than this have been managed with success, though it may be necessary to assess the requirement of some other treatment form in addition to the Gamma Knife. Both partial surgical resection and embolization are alternatives which would have to be considered.

Fig. 10.4
(a) Angiography is vital because it gives resolution in time. The pictures show the nidus (arrow) in an early picture in the angiogram sequence. (b) The nidus soon becomes obscured by the draining veins (arrows), seen here on a slightly later picture in the angiogram sequence. It is not necessary to include these veins in the radiation field. Thus the field is kept to a minimum by restricting it to conform with an accurate delineation of the nidus as shown on the earlier pictures in the angiogram series

Acceptance of Patient for Treatment

If a patient meets the criteria outlined in the previous paragraph then that patient may be accepted for treatment. However, it seems appropriate to include the patients themselves in the decision-making process. What will suit one will not necessarily suit another. This is an area which arouses passions and leads to controversy. However, with the modern development of patient associations, it seems wise to accept that changes are under way and that the specialist's paternalism of yesteryear is perhaps no longer appropriate.

The treatment is, as has been stated, prophylactic. It is not without risk but it is a statistically lesser risk than that of the untreated disease. However, the risks of treatment as far as we know are the same irrespective of the mode of presentation of the patient. Again, some patients tolerate the presence of an intracranial AVM badly, particularly if they have been alarmed by a referring physician, who has been impressed by the dramatic looking X-rays. If surgery is a viable alternative, the patients must be made aware of this. Modern AVM grading systems and microsurgery, have resulted in a great improvement in the results of open surgery. Unfortunately, senior non surgeon authors who write leading articles and reviews, often refer to a literature from 10 to 20 years ago where the results were very much worse than those reported today. This must also be taken into account, as such articles are important in forming opinion. The fact is, that for surgically accessible malformations of the size appropriate for Gamma Knife treatment, the mortality is today virtually 0%. However, there is the morbidity and cosmetic defect associated with surgery to be considered. How relevant this is, in the author's experience varies considerably from patient to patient. Most appear to prefer the Gamma Knife when offered an alternative. However, the choice is more apparent than real, because the majority of patients referred have malformations which are not appropriate for surgery, usually because of their location.

Follow-up

The patients must be followed until the AVM is declared angiographically obliterated. However, repeated angiograms are not necessary today. Ideally an MRI examination every 6 months will suffice with an angiogram either when the AVM is invisible on the MRI or at 2 years, which ever occurs first. Nonetheless, some young active patients find the two year wait intolerable and request angiographic control at 1 year. Since the chance of obliteration at one year is about

50%, this seems reasonable for this restricted patient group. This particular patient category is also extremely appreciative of being shown consideration and declared fit a year earlier. Many are involved in athletics, where taking advantage of their youth is of course of the essence.

It also seems to be a characteristic of many patients with AVMs that they remain nervous, even after the AVM is no longer present. This is perhaps understandable, since they will, even in the best situation have suffered the feeling that they have a bomb in their heads, that can explode at any minute, for a minimum of over 12 months. Thus, it is incumbent on the physician to show these unfortunate people the greatest possible patience when they ring up or write again and again for reassurance that they really are fit and that the danger is past. In the author's view referral of AVM patients to a psychiatrist at any stage of their treatment is inappropriate, unless they themselves wish it. It cannot be pathological to be anxious about having an AVM in one's head and thus such anxiety is a more sensibly alleviated by reassurance than therapy. However, if the patient wants therapy that too can be an effective form of reassurance. However, if the patient does not want therapy, insisting on it can greatly undermine that patient's relationship with the physician responsible for treating the AVM.

Another often repeated question from AVM patients is to what extent they must restrict their lives and for how long. This seems to be particularly difficult to answer with any degree of confidence, since the relationship of AVM bleeding to short term energy consuming pursuits is not well defined. However, the restrictions that are to be followed should logically continue until such time as angiographic obliteration of the AVM has been demonstrated.

It seems to the author unreasonable as well as impossible to restrict the every day activities of children, especially pre-school children. However, for school children, avoidance of gymnastics and competitive physical games and sports is advised. These activities carry an unavoidable risk of increasing the load on the circulation, and in many competitive physical games also increase the risk of direct damage to the head. On the other hand walking, jogging, skating, swimming (providing epilepsy is not a contraindication) and in northern regions skiing over familiar terrain all seem reasonable. Some of these activities can be characterized as non-competitive sports and these should only be undertaken in the company of an informed adult companion.

Patients are often concerned about how restricted their love making should be. Again there is no simple answer. Advice along the lines of being gentle rather passionate may be helpful. It seems an unreasonable imposition to forbid love-making totally. While it is known that a certain number of bleeds occur during sexual activity a far greater number occur at other times. Moreover, a blood pressure rise occurs every time one passes urine, yet this activity cannot be forbidden. To put the problem in perspective, despite the limitations that must pertain to information derived from a single case, the author would like to mention the experience of a particular patient. This lady had been an air stewardess and suffered a cabin decompression at high altitude. She had subsequently changed work and gone to work with horses. During this period of her life she was both kicked on the head and suffered falls on the head. Her malformation finally bled the day after an altercation with a potential mother-in-law. It is really very difficult to assess the dangers of specific activities and while total inactivity is *probably* the safest thing even this is uncertain. In the meanwhile, total compulsory inactivity, if enforced rigourously against the patient's wishes can have serious psychological effects, which can outlast the successful treatment of the malformation, irrespective of the method of treatment involved. However, the inherent delay involved in Gamma Knife treatment makes the careful evaluation of the advice given to patients even more important than is otherwise the case for those suffering from a cerebral arteriovenous malformation.

Cavernous Haemangiomas

Cavernous haemangiomas, otherwise known as cavernomas are one of the diagnoses that is included in the abbreviation AOVM. This stands for **A**ngiographically **O**ccult **V**ascular **M**alformation. Cavernomas consist of cavernous, sinusoidal vascular spaces, lined with venous endothelium and without cerebral parenchyma in the interstices. They have been the subject of much discussion, but statistical analysis of their natural history is a development of only the last few years. The improved localization and diagnostic certainty, following the use of MRI for these lesions has resulted in a reappraisal of their suitability for radiosurgery. They bleed but these bleeds are seldom lethal. The rebleed rate seems to be well under 1% a year. Many of the cavernomas, which are placed in the cerebral hemispheres cause epilepsy. Using the MRI, these lesions make splendid targets and there is some evidence that Gamma Knife surgery will improve associated epilepsy. However, it is very difficult to prove occlusion of a cavernoma. This is because the typical MRI appearances are produced by haemosiderin deposits. Even if the lesion should occlude, the haemosiderin does not disappear. Thus, the MRI appearances remain unchanged, irrespective of the effect of treatment on the risk of bleeding. Since they are invisible on angiography and less well seen on CT, there is no help to obtained with these investigations. It seems sensible to advise Gamma Knife treatment for well defined lesions, away from the neuraxis. However, it is at present conventional to be very careful in advising the Gamma Knife for brain stem lesions. The doses, which are thought to be required would very probably represent as great or greater danger to the patient as the untreated disease; at least in many cases.

Suggested Further Reading

1. Backlund E-O (1979) Stereotactic radiosurgery in intracranial tumours and vascular malformations. In: Krayenbühl H, et al (eds) Advances and technical standards in neurosurgery, vol 6. Springer, Wien New York, pp 1–37
2. Curling OD, Kelly DL, Elster AD, Craven TE (1991) An analysis of the natural history of cavernous angiomas. J Neurosurg 75: 702–708
3. Garner TB, Curling OD, Kelly DL, Laster DW (1991) The natural history of intracranial venous angiomas. J Neurosurg 75: 715–722
4. Kihlström L, Lindquist C, Adler J, Collins P, Karlsson B (1992) Histological studies of gamma knife lesions in normal and hyper-cholesterolemic rabbits. In: Steiner L (ed) Radiosurgery, baselines and trends. Raven Press, New York, pp 111–121
5. Lindquist C, Steiner L (1988) Stereotactic radiosurgical treatment of arteriovenous malformations. In: Lunsford LD (ed) Modern stereo-tactic neurosurgery. Martinus Nijhoff Publishing, Boston, pp 491–505
6. Lunsford LD, Kondziolka D, Flickinger JC, Bissonette DJ, Jungreis CA, Haitz AH, Horton JA, Coffey RJ (1991) Stereotactic radiosurgery for arteriovenous malformations of the brain. J Neurosurg 75: 512–524
7. Nilsson A, Wennerstrand J, Leksell D, Backlund E-O (1978) Stereotactic gamma irradiation of the basilar artery in cat. Acta Radiol 17: 150–160
8. Ogilvy CS (1990) Radiation therapy for arteriovenous malformations. Neursurgery 26: 725–735
9. Robinson JH, Awad IA, Little JR (1991) Natural history of the cavernous angioma. J Neurosurg 75: 709–714
10. Scott BM, Barnes P, Kupsky W, Adelman LS (1992) Cavernous angiomas of the central nervous system in children. J Neurosurg 76: 38–46
11. Steiner L (1985) Radiosurgery in cerebral arteriovenous malformations. In: Fein HM, Flamm ES (eds) Cerebrovascular surgery. Springer, Wien New York, pp 1161–1215
12. Steiner L, Lindquist C, Adler JR, Torner JC, Steiner M (1992) Clinical outcome of radiosurgery for cerebral arteriovenous malformations. J Neurosurg 77: 1–8

11. Acoustic Schwannomas

Introduction

After arteriovenous malformations, acoustic schwannomas have been the most rewarding condition to treat in the Gamma Knife, which is reflected by the large number of patients treated world wide. Nonetheless, the correct way to treat this difficult condition is a matter of debate. Surgery may be performed, by the translabyrinthine approach by an ENT surgeon. It may also be performed through the posterior fossa by a neurosurgeon. Ideally both forms of surgery may be available with collaboration between the specialities involved. A third operation, through the middle fossa has been reserved for hearing preservation in small, primarily intracanalicular tumours. Another quite different form of treatment is of course that provided by the Gamma Knife. In 1991 two major assessments of acoustic treatment were presented. One was an article summing up the results of a century of treatment. The other was the first international acoustic "neuroma" conference held in Copenhagen, where top expertise from all relevant specialities and most parts of the world were present. On the basis of data provided from these two sources, it is fair to say that there is a considerable range of opinion on the subject of what constitutes optimal treatment. As far as the Gamma Knife is concerned, it has been and indeed to some extent remains the convention to say, that the Gamma Knife is appropriate for the treatment of patients who are a high surgical risk, of advanced age or who refuse open surgery for whatever reason. The correctness of this view will be debated further a little later on. As a neurosurgeon the author would like to deplore the use of the term acoustic neuroma for this tumour. This is an etymological misuse which has become widespread. If scientific prose is to be precise then the terminology must be used precisely. The word neuroma indicates that the tumour's cell of origin is a nerve cell which is not the case. The tumours are thought to be derived either from Schwann cells or neurilemmal cells and are thus variously referred to as schwannomas, neurinomas or a neurilemmomas. The use of any of these three terms can be defended: but neuroma is nosologically incorrect. In this text schwannoma is preferred.

Diagnosis

The majority of patients present with progressive unilateral hearing loss, though a substantial minority present with a wide variety of symptoms related, to malfunction of any or all of the structures in the cerebello-pontine angle. These symptoms include tinnitus, vertigo, ataxia, facial numbness and mild facial palsy. Moreover, with bilateral tumours, the symptoms may develop bilaterally. Thus, diagnosis depends on a high index of suspicion. When the diagnosis is suspected, neuroradiological and neuro-otological studies are performed. The classic neuroradiological findings are a cerebello-pontine angle tumour, with on CT a moderate degree of contrast enhancement which is not necessarily uniform. Moreover, the internal auditory canal should be widened. None of these features are however pathognomonic for the acoustic schwannoma.

The neuro-otological work up includes brain stem auditory evoked response, electronystagmography and caloric testing as well as audiological examination. The degree of diagnostic certainty with positive MRI, CT and neuro-otological findings is considered to be high enough to permit treatment even without biopsy, something which is unavoidable if Gamma Knife surgery is the primary treatment used.

Results of Different Forms of Treatment

In a modern review of the management of acoustic schwannomas, a summary of the results of treatment over the last 15 years shows an overall total tumour removal in all forms of surgery of 95.5%. This is equivalent to 95% for the suboccipital approach, 96.8% for the translabyrinthine and 99.4% for the middle fossa approach. These results were associated with an overall mortality of 1.8% divided as follows; 2% for the suboccipital approach, 1.7% for the trans-labyrinthine approach and 0.8% for the middle fossa approach. Moreover, preservation of the facial nerve was 73% for the suboccipital approach, 75% for the translabyrinthine approach and 98% for the middle fossa approach.

Results with the Gamma Knife from Stockholm, quoted in the same article gave control of the tumour in 91% of patients with unilateral tumours and facial nerve preservation in 100%. Moreover, while facial nerve deficits and in some cases even paralyses were seen, these were invariably temporary. Nonetheless, there was a risk of some degree of facial deficit after a total paralysis though this was never severe enough to be disfiguring. The group in Pittsburgh have reported a better tumour control rate (98%), with a rather higher dose than that which is current in Europe. However, this group has also observed a temporary facial palsy rate of 35%, in its material. They have subsequently reduced their dose. Our own results show an early success rate of 92% with 12% temporary slight facial palsy. There is on-going work at the three centres in an attempt to define optimal dose-planning. Nonetheless, these results with their associated zero mortality are very attractive.

Aims of Treatment

These are fairly simple. As with any tumour, the main aim of treatment is to prevent further growth. If in addition there is reduction in tumour volume this is an additional bonus. Furthermore, the therapeutic success should be achieved with no disturbance of facial or acoustic function. While this latter aim may be impossible to achieve, it nonetheless remains an aim of treatment.

Choice of Treatment or Treatment of Choice

It is important to underline that this section relates to unilateral tumours of 2.5 to 3 cm maximum diameter and not to larger tumours, where the role of Gamma Knife surgery is at present uncertain. Moreover, it does not apply to patients with bilateral acoustic tumours (usually with neurofibromatosis) for reasons to be given later. In addition, it is important to emphasise that the views expressed here are the author's personal views based on the results of colleagues in Sweden and the USA and on his own short term experience over four years. They are not shared by many colleagues at the present time. The author would suggest that even with today's results the Gamma Knife may be considered the treatment of choice for the primary management of these small unilateral tumours. The basis for this view is as follows. Firstly, it must be remembered that much of the anxiety that relates to the condition of acoustic schwannoma is an anachronism, inherited from a time where relatively primitive clinical and radiological techniques often delayed diagnosis until brain stem compression was a major immediate concern. This is not the case with tumours of the size that are under discussion here. Moreover, while the question "Do all small acoustic neurinomas continue to grow?" remains unanswered it seems likely that they do. At all events, regular follow up with MRI and gadolinium contrast will enable the individual patient to be observed safely. Thus, considering only the neoplastic threat posed by an acoustic neurinomas, there is no rush to remove them immediately by surgery. In consequence, observation alone or observation following Gamma Knife therapy may be considered responsible management alternatives. Moreover, at the time of diagnosis most patients have mild to moderate symptoms, mainly related to hearing. However, while retention of useful hearing is only found in a minority of patients at presentation today, with improving diagnostic methods, the proportion with useful hearing at presentation is likely to increase, in the future. Improvement of hearing following treatment is extremely rare even for those with preserved useful hearing. Nonetheless, preservation of useful hearing is a high priority indicating that when this is a consideration, treatment should not be delayed too long. At the present time, the Gamma Knife offers amongst the most satisfactory results in relation to preservation of hearing. In respect of another common symptom, tinnitus, no form of treatment available today has a **predictable** useful effect. Thus, the treatment the patient will receive, whether it be microsurgery, Gamma Knife surgery or just observation, cannot be expected to have any reliable effect on improving this symptom. Thus, with the possible exception of preserving useful hearing, treatment is unlikely to improve the patient's symptoms greatly. Nonetheless, early treatment with a small tumour has been demonstrated time and again to give better results than treatment of a large tumour. Thus, while it is acceptable to wait to see if a small tumour is growing or not, one should not wait too long, or have too long intervals between the observations on which the assessment is to be based.

The above discussion indicates that there is a choice of available treatments and a choice of timing of treatment, depending on the level of preserved hearing and the size of the tumour. Nonetheless, most tumours presenting today do not represent a short term threat to the brain stem and as such do not represent a short term threat to life. Thus, treatment is, *at the time it is carried out*, essentially prophylactic and its results should not be worse than the disease that is being treated. In this context there is a very real risk with open surgery to acoustic neurinomas, even if the frequency of complications of open surgery is low. Firstly, there is a small possibility for damage to the anterior superior cerebellar artery with a brain stem infarct as the result. Even in the best hands surgical treatment of small tumours still has a mortality. Secondly, there is a measurable risk for developing a disfiguring facial palsy following surgery. Thirdly, there is the disruption of life involved in a hospital stay of several days at best and weeks if a complication occurs. Moreover, there is a measurable risk of CSF leakage which may or may not require drainage procedures and may even require an additional operation to repair the leak. Finally, there is the risk of surgical infection. If all this is borne in mind and if it is accepted that a facial disfigurement is a social catastrophe, while unilateral deafness is merely a social inconvenience, then the argument for non-surgical treatment, if such be available becomes compelling. In the face of the results outlined above, it is suggested that the Gamma Knife may well be the treatment of choice rather than a choice of treatment. An effective counter to this argument would be that in the event of treatment failure, the operative conditions would be worsened by the previous Gamma Knife surgery. And indeed, there is some anecdotal evidence to this effect. However, the cases that have been described as difficult to operate have not been associated with any complications of note, for the patient. Moreover, these cases hail from a time when much higher doses were used than is now the case. Anyone who has operated in an intracranial field, which has been subject to a conventional fractionated irradiation with a total dose of 50 to 60 Gy will be familiar with the grey, tough but friable tissue which makes surgery more difficult. However, the author has operated on four intracranial lesions, which have been subject to prior Gamma Knife treatment, where the dose beyond the tumour has not exceeded 12 Gy and the dose to the tumour has not exceeded 40 Gy. These four tumours consisted of three pituitary adenomas and one acoustic schwannoma. The findings were uniform and monotonous. The edge of the tumour was in no way peculiar for its location. No extra adhesions were observed and the tumours were rather avascular and easy to remove. Thus, modern low dose radiosurgery would not appear to make subsequent surgery more difficult. If anything, rather the reverse is the case.

Changes in Acoustic Neurinomas Following Radiation

These have for the most part been observed at X-ray rather than in the laboratory. One patient was reported by Norén in 1983 where the post-mortem findings were observed, after the patient died of pneumonia. There was a sharply defined volume of necrosis with an edge corresponding to the 50% isodose line, in this case corresponding to 50 Gy. It is of interest that this dose is much higher than that which is considered appropriate now. The radiological findings following treatment will be described under the section related to follow-up.

Patient Assessment for Gamma Knife Surgery

Since at the present time there is no consensus as to the treatment of choice for acoustic neurinomas, it seems sensible to assess each case by means of a team consisting of at least an otolaryngologist, a neurosurgeon and a neuroradiologist. The neurosurgeon should also have expertise in the use of the Gamma Knife. With such an arrangement as this, it is possible to offer the patient a choice of treatments, best suited to that patient's individual needs. Some want to be quit their tumour at once and do not tolerate the idea of the presence of an inactivated tumour. However, experience amongst surgeons, with whom the author has spoken gives the impression that it is often the surgeon rather than the patient who does not like the idea of an inactivated retained tumour. An acoustic team had been established in Bergen before the Gamma Knife was installed. Since otolaryngologists are inevitably involved in the investigation of all of these patients it made sense for an otolaryngologist to be the chairman of the acoustic group. Even so, since the Gamma Knife was installed, approximately two patients have been treated radiosurgically for every patient who has been treated surgically. The team makes its decisions on the basis of the clinical presentation, the neuro-otological findings and the radiological findings. MRI is especially useful for intracanalicular tumours, while CT remains the best examination to determine the bony changes associated with these tumours.

Acceptance of Patient for Treatment

After the patient has been allocated for treatment he/she is informed of which type of treatment is recommended and if it is agreeable then an appointment is made. On admission, careful explanations are made relating to the effects of the treatment on hearing on facial function and on the tumour. The patients with useful hearing are informed that there can be no guarantee that hearing will be retained, though deterioration can be delayed and may not occur. The delay in deterioration is anything from a few months to five years in most cases. Patients are moreover informed that facial nerve function is temporarily affected in a small proportion of cases and that this occurs usually after a delay of six to eighteen months. If the tumour is entirely intracanalicular or has a volume of less than 1000 mm^3, then the chance of even temporary facial nerve damage is effectively nil. The chance of deterioration in hearing is also much reduced with such small tumours, though at the present time this can not be quantified.

Patients with tinnitus or vertigo are informed that they cannot expect any immediate effect upon these symptoms. Indeed, the effect of the Gamma Knife on these symptoms is at best uncertain even in the long term. However, this is the same for all types of treatment. The patient must also be informed of the long term follow-up over many years and that the aim of treatment is to prevent the tumour growing rather than to remove it. This is often a cause of some confusion. There is some uncertainty as to how long follow-up is necessary to determine if a cure has been achieved, within the frame of reference of the technique. The final answer to this question is at present unknown. However, Norén has suggested that there is very little if any likelihood of late regrowth if no growth has been observed over a five year period. At present Norén is the only worker with a sufficiently long term follow-up to be able to have a view on this topic.

Dose

This has varied over the last 15 years. There is broad agreement today that the dose to the edge of the tumour should be about 15 Gy. Though the dose will be lower for patients with only one hearing ear, for example in cases of neurofibromatosis.

Patient Follow-up

Because of the uncertainties related above, follow-up in Bergen is planned at the present time to continue for at least eight years. It must consist of clinical, neuro-otological and radiological examinations. CT or preferably MRI with appropriate intravenous contrast is carried out at six months, one year, two years, four years and eight years and ten years. The clinical picture is not expected to change much. However, as mentioned above, in a few patients a temporary facial palsy may develop at six to eighteen months after treatment. This must be registered and graded and followed until it stabilises, usually with disappearance of the paresis. A similar practice must be followed with the development of trigeminal sensory deficit. This would seem to be rather more recalcitrant and persistent, as is common with other sensory deficits. Hearing is followed meticulously. The X-rays will often show a region of diminished contrast enhancement in the centre of the tumour at the first examination at six months (Fig. 11.1). This has a tendency to disappear again after about a year. It is taken to indicate the presence of radionecrosis, though this is not confirmed. It is also taken to indicate a positive response to treatment. Some tumours will then show signs of shrinkage. In this context, uniformity of radiological technique between referring department and treating department makes comparative size measurements much simpler. It would appear that the longer the follow-up period the greater the proportion of tumours which will show signs of shrinkage. While genuine growth must be the cause for alternative treatment it is important to be aware of the possibility, in the early stages, of some reversible tumour swelling which subsequently regresses. In general, this is suspected if lack of contrast enhancement is found together with a slight increase in tumour volume. Since MRI techniques have been used, another source of false tumour growth is due to temporary reversible oedema, in the neighbouring brain stem extending the region of contrast enhancement and mimicking tumour growth. This phenomenon is well described by the Pittsburgh group. Thus, it is not necessarily desirable to resect a tumour if a single follow up examination indicates slight increase in tumour volume. It is more appropriate to wait until the next examination, providing there is not additional clinical evidence that gives cause for concern.

Patients with Bilateral Tumours

These tumours are a problem for all forms of treatment. They are usually a component of neurofibromatosis 2, shortened hereafter to NF2. These tumours are difficult in a variety of ways. Firstly, because they are often bilateral the consequences for hearing are far more important than with unilateral tumours. Secondly, they often have a tougher consistency, making conventional surgery more difficult. Thirdly, the nerves whose function should be preserved pass through the tumour rather than over the surface, as with the usual form of schwannoma. This places the facial and acoustic nerves at risk both during open surgery and following Gamma Knife treatment. Dose-planning is obviously a more difficult undertaking if the site of the structures to be avoided is uncertain and when they are more likely to be in the centre rather than at the edge of the tumour being treated. All these factors result in treatment for this category being less satisfactory in respect of tumour control. However, they continue to be referred, often following surgery to the side with least hearing, in an attempt to stop tumour growth while preserving what hearing is retained, on the contralateral side. In general, the dose given is lower than for the unilateral schwannoma and this may play a part in the poorer control of tumour growth, which is between 65 % and 70 % in Norén's material. However, this technique has retained unchanged hearing at one year after treatment in 36 % of patients as opposed to 24 % of unilateral tumours. Thus, it probably is the best available

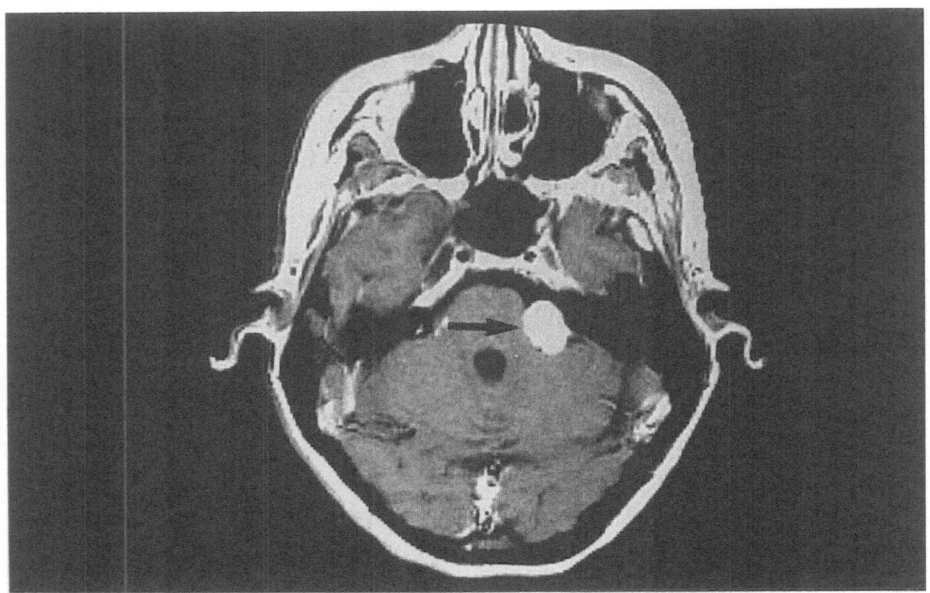

Fig. 11.1. a

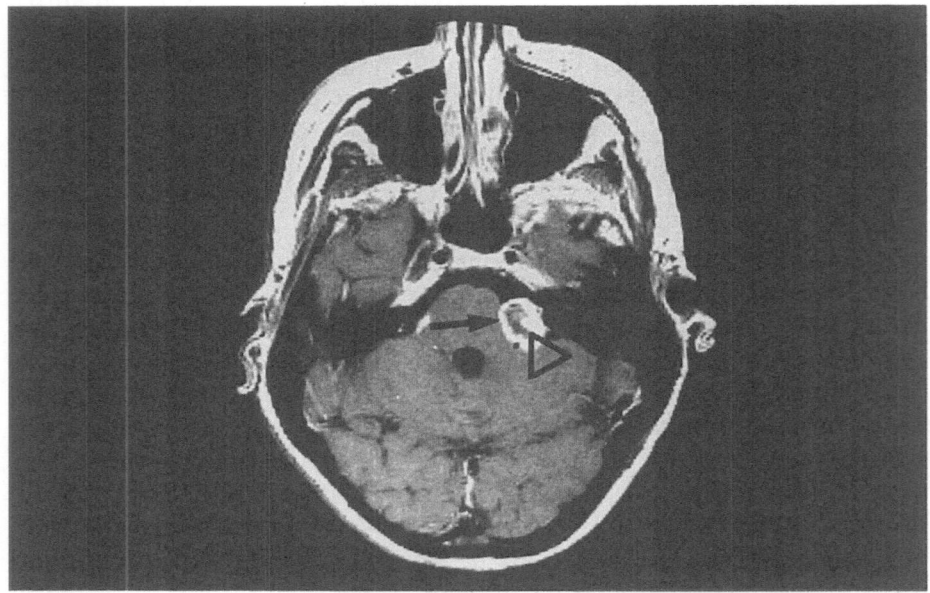

Fig. 11.1. b

treatment today to preserve hearing in a patient afflicted with NF2, where previous treatment or the disease has destroyed hearing in the contralateral ear.

Suggested Further Reading

1. Brackman D, Kwartler JA (1990) Treatment of acoustic tumours with radiotherapy. Arch Otolaryngol Head Neck Surg 116: 161–162
2. Flickinger JC, Lunsford LD, Coffey RJ, Linskey ME, Bissonette DJ, Maitz AH, Kondziolka D (1991) Radiosurgery of acoustic neurinomas. Cancer 67: 345–353
3. Ganz JC, Myrseth JR, Thorsen F, Backlund E-O (1992) Acoustic schwannoma: early results of radiosurgical treatment. Copenhagen Acoustic Neuroma Conference Proceedings. Kugler Publ., Amsterdam, pp 301–304
4. Linskey ME, Lunsford LD, Flickinger JC (1990) Radiosurgery of acoustic neurinomas: early experience. Neurosurgery 26: 736–745
5. Lunsford LD, Flickinger JC, Coffey RJ (1990) Stereotactic Gamma Knife radiosurgery: initial North American experience in 207 patients. Arch Neurol 47: 169–175
6. Norén G, Greitz D, Hirsch A, Lase I (1992) Gamma Knife radiosurgery in acoustic neurinomas. Copenhagen Acoustic Neuroma Conference Proceedings. Kugler Publ., Amsterdam, pp 293–296
7. Norén G, Arndt J, Hindmarsh T (1983) Stereotactic radiosurgery in cases of acoustic neurinoma: further experiences. Neurosurgery 13: 12–22
8. Norén G, Arnst J, Hindmarsh T, Hirsch A. (1988) Stereotactic radio-surgical treatment of acoustic neurinomas. In: Lunsford LD (ed) Modern stereotactic neurosurgery. Martinus Nijhoff Publishing, Boston, pp 481–489

Fig. 11.1
(a) The MRI appearance (with gadolinium) of a left sided acoustic neurinoma (arrow). (b) An equivalent MRI examination of the same patient six months after treatment. The tumour (arrow) is the same size but it shows the characteristic loss of contrast enhancement (open arrowhead). The pathological basis of this change is unclear at the present time. It probably does not indicate necrosis, for two reasons. Firstly, it disappears again over the next few months in many cases. Secondly, in one patient, previously treated in the Gamma Knife, and who had shown this change, who was then explored surgically by the author, no macroscopic or microscopic necrosis could be seen. However, whatever the nature of this radiological change, it seems to be a fairly reliable indicator of a successful result in the long term

12. Gamma Knife Applications in and around the Pituitary Fossa

Introduction

The pituitary region was one of the first regions to be attacked therapeutically, with open stereotactic technique. This is because the soft tissues have a relatively fixed relationship to the base of the skull and could be outlined with adequate detail, using different forms of air study. In consequence, adequate localization of the pituitary region, for puncture of cysts located in that region, was available at an early date. Thus, Leksell's very first patient in his first frame was treated with instillation of ^{32}P into a craniopharyngioma cyst; as early as 1948. Moreover, Wycis had treated a cystic pituitary adenoma with intracavity ^{32}P in 1954. However, the use of this isotope against pituitary adenomas never really caught on. It was Backlund who first began the systematic treatment of tumours of this region with stereotactic technique, using the intracavity instillation of Yttrium-90 to treat craniopharyngioma cysts. This treatment was to prove highly successful. Thus, it was natural that one of Backlund's patients should be the first candidate for Gamma Knife surgery, as mentioned in chapter 6.

The pituitary region in many ways provides the greatest intellectual challenge of any region which is treated by radiosurgery. Firstly, there is the usual treatment of neoplasia, as exemplified by the use of radiosurgery for the solid component of craniopharyngiomas. Then there is the use of the Gamma Knife to perform hypophysectomy. Indeed, hypophysectomy represents quite a departure from all that has been described so far in this book, in that it involves the radiation destruction of non-cerebral normal tissue. Thus, it was necessary to define the radiosensitivity of that tissue, that is to say the anterior pituitary lobe. Because that is what the Gamma Knife offered; the possibility of performing anterior lobe ablation, without opening the skull. This is a particularly attractive therapeutic alternative for the patients for whom such treatment is relevant, since they are in the main in poor condition, because of advanced cancer. This will be discussed later.

It was a natural progression to think that if anterior lobe pituitary ablation were possible, then it should be possible to treat pituitary adenomas. Initially a variety of adenomas of various sizes were treated. Though great care had to be exercised, with the available X-ray techniques, because of the risks of radiation damage to the optic nerves, the chiasm and/or the hypothalamus, with larger tumours. The systematic treatment of a pituitary endocrinopathy was another new departure. The aim was to attempt to eradicate the pituitary adenoma and at the same time correct the endocrinopathy, without producing pituitary failure. Cushing's adenomas were chosen as they are in general the smallest of pituitary adenomas and this facilitates the delivery of the necessary high dose to the tumour, while keeping the dose to the visual pathways and the hypothalamus low. However, even today, it is difficult to visualize these adenomas so that using CT scans the tumour can be seen in only about 50 % of cases. However, with MRI studies using gadolinium the number of cases has been increased to as high as 90 %.

Thus, in the pituitary region, all the principles of Gamma Knife treatment were concentrated in a new way. The radiosensitivity of a new tissue had to be determined and the differential sensitivity between adenoma tissue and anterior lobe tissue had to be defined. The avoidance of particularly sensitive regions, the hypothalamus and the optic pathways was crucial to success. Finally, radiation is to be used not only for the control of neoplasia but also as a selective chemical regulator. These were great challenges, and as will be seen have tested Gamma Knife radiosurgical technique to the limit.

Hypophysectomy for Cancer Pain

Choice of Treatment

This was a development of the use of hypophysectomy to control hormone dependent cancers, in particular breast cancer and prostate cancer. On the whole, the operation had been a disappointment. It was moreover viewed as a major extra burden for an extremely sick patient. The indications had been restricted to patients with bone pain from bone metastases, who could thereby experience relief of this pain. The only published series of Gamma Knife pituitary ablation consisted of 8 patients with breast cancer and bone metastases. 4 of these died within 8 months. The remaining surviving 4 experienced increased well being and reduction in bone pain.

Radiosensitivity of the Anterior Hypophysis

The aim of treatment was clearly to produce an early lesion and this could be achieved consistently with a maximum dose of 200 to 250 Gy or 100 to 125 Gy to the edge of the anterior hypophysis. The required dose to produce a speedy response is still not certain but a dose of between 120 to 180 Gy is thought likely to produce an acute lesion, with anterior pituitary ablation within 3 to 8 weeks. The Gamma Knife permits the construction of dose-plans with an adequate dose to the anterior pituitary, while sparing the visual system from a dose of more than 20 Gy. This is a rather high dose but thought to be acceptable in view of the desperate nature of the condition being treated.

At the time of writing there is not much more information about this patient category. In the opinion of the author, this represents an attractive, underused therapeutic alternative for a moderate number of desperately suffering patients and it is to be hoped that more patients will be referred in the future. In this particular situation, there is a particularly stringent requirement for the Gamma Knife department to follow up all patients and to obtain an autopsy with the intact pituitary. In this way, a more accurate measurement of pituitary radiosensitivity may be obtained and treatment may be thus made more selective and sophisticated.

Cushing's Disease

Radiosensitivity of Pituitary Adenoma Tissue

It has been a contention of those engaged in the treatment of Cushing's disease that a dose of 70 Gy, accurately delivered to the adenoma will produce tumour necrosis. This is clearly a much lower dose than that required to produce an acute anterior lobe ablation. Moreover, if the development of necrosis parallels the endocrinological return to normal then this necrosis takes many months to 3 years to develop as opposed to a few weeks. Whatever central tumour dose is chosen, it has been accepted, as a condition of treatment, that the visual pathways should not receive more than 10 Gy.

Diagnosis

As may be inferred from the introductory paragraphs, diagnosis is not just a simple matter of radiological identification. More detailed accounts of the clinical picture and evaluation of patients with Cushing's syndrome are to be found in general textbooks of medicine. However, it is important do distinguish between the terms Cushing's syndrome, which describes the clinical complex irrespective of cause and Cushing's disease, which indicates a primary pituitary origin. It is obviously the latter in which the Gamma Knife user is interested. The main differential diagnosis is from primary adrenal causes of Cushing's syndrome, – hyperplasia, adenoma or carcinoma – and ectopic ACTH production from, for example, a bronchogenic carcinoma. This latter while well defined and recognized is even rarer than Cushing's disease, which has an incidence of about 2.3 per million. To distinguish between the different kinds of Cushing's syndrome, in the presence of a normal radiological examination of the pituitary fossa is not easy. This is because of the small amounts, general variability and diurnal variation of the hormones that are measured. However, with the use of a battery of tests the endocrinologist, in most cases will come to a correct diagnosis.

Aims of Treatment

The aims of treatment in Cushing's disease are to remove the pituitary adenoma without producing additional morbidity or mortality. In particular pituitary failure should be avoided. In children particular attention must be focused on the effects of treatment on growth. Cushing's disease in childhood occurs sufficiently often that special consideration must be taken for this group of patients.

Results of Different Forms of Treatment

At the time that Gamma Knife surgery was started, the conventional surgical removal of adenomas causing Cushing's disease, with correction of the endocrinopathy, was unsatisfactory. However, greater experience with the microsurgical techniques and radio-immunoassay and MRI studies have resulted in improved surgical results. Thus, a surgical cure can be expected today in between 70% to 80% of cases, with only slight risk of compromising remaining pituitary function. The results of Gamma Knife surgery on Cushing's disease has on the whole been disappointing, when compared with microsurgery, with success at one session being achieved in only about 50% of cases. However, if the dose is repeated – to a total of from 2 to 4 doses – it has been possible to obtain remission in 76% of cases. However, this takes from 1 to 3 years. On the other hand the results of microsurgery are immediate and not delayed. This is important in such a debilitating condition as Cushing's disease, which untreated is reported to have a life expectancy of less than five years. However, the proponents of radiosurgical treatment claim that these poor early results are a consequence of the limitations of localization, at the time the studies were carried out. That is to say, the results referred to in this section are taken from studies which were performed before CT or MRI were available. This is a reasonable objection. Even so, if radiosurgical treatment is to be repeated more than once then the risk of pituitary failure, due to summation of dose to the normal pituitary is bound to be more prevalent. On the other hand, the results from Stockholm with children was more encouraging with a remission in seven of eight children following a single treatment. Nonetheless, all these children suffered some degree of growth rate reduction, beginning in the second year after treatment.

In Bergen we have observed a number of patients whose post treatment CT examinations indicate that tumour necrosis has been achieved without adequate control of the endocrinopathy. In this situation localization has not been the issue. On the other hand, the dose in these patients has been rather low. Nonetheless, of four patients with more than 18 months follow up, two have been cured by the Gamma Knife and two have been cured by a subsequent micro-surgical operation, making the cure rate for this small series 100 %.

Choice of Treatment

This is today a matter of controversy. It is the author's belief that the Gamma Knife, on the basis of current published evidence is not the best primary treatment for Cushing's disease because of the technical problems outlined in the previous section. However, on the basis of his own experience together with some newer findings in Stockholm, the time appears right for a reappraisal. It is necessary to perform a prospective study, using the Gamma Knife only for tumours adequately localized in one way or another. These should be followed up for only six months and if the result is inadequate, the patients should then be offered a microsurgical operation. It is inappropriate to repeat treatment in the Gamma Knife for such a serious disease, with all the delays concomitant upon such a strategy. The evidence noted above suggests that such a study could document a very promising therapeutic alternative. However, the highest possible dose consistent with safety of the optic apparatus, will be crucial for success. So, the last word has certainly not been said in this matter. However, as matters stand today, even if the Gamma Knife is not the preferred primary treatment it certainly has a role to play as an adjunct to treatment, if microsurgery seemingly removes the tumour but all the same does not produce an endocrinological cure.

Patient Follow-up

Patients who are treated in the Gamma Knife will need to be followed for at least five years to assess both the change in the endocrinopathy and to exclude the development of pituitary failure. The delayed nature of the response to treatment means that both the patient and his/her referring physician require constant reassurance. There is often a marked pressure to proceed to surgery before the treatment has had a fair chance to work. Visual function should also be checked annually, even if the patient does not complain of any visual disturbance.

Prolactinoma

Prolactinomas are in general best treated with dopamine agonists, of which bromocriptine is the most frequently used. However, some patients tolerate medical treatment badly and they can reasonably be treated with radiosurgery. However, there is some evidence that prolactinomas are more radioresistant than ACTH producing adenomas, so that transsphenoidal surgery is often a more attractive second choice of treatment for these patients.

However, prolactinomas are, of all primary hypersecreting adenomas, most likely to become large and invasive. While it is possible to follow a tumour into the cavernous sinus, as described by Fahlbusch, this is at best an uncertain procedure, because the region is necessarily difficult to visualize at operation and because of the importance of not damaging the nerves and vessels. Thus, it is the most likely place for residual tumour to be found, following surgery.

Fortunately, with one proviso, a tumour in the cavernous sinus region is a easier to treat in the Gamma Knife than a tumour in the pituitary fossa, because it is usually at a greater distance from the visual pathways. The proviso is, of course, that the tumour may be adequately defined radiologically. Moreover, the carotid artery and the nerves in the wall of the cavernous sinus seem to tolerate radiosurgical lesions well. Thus, residual tumour in this region, particularly if it shows signs of growth can very reasonably be controlled by a Gamma Knife lesion. In addition, there is a good chance of improving the endocrinopathy, though it will be difficult to abolish it. According to the experience in Bergen, based on three patients the endocrinopathy improved in all. In two the prolactin levels, previously high despite bromocriptine became normal on the same or a reduced dose of bromocriptine. The third patient achieved improvement though not abolition of the hyperprolactinaemia. However, in these cases the main aim has been to stop the spread of more than usually aggressive tumour growth. In this aim, success has been achieved in all three patients over an observation period of two to three years.

Acromegaly

The same indications that are described for prolactinomas apply also to patients with acromegaly, with residual tumour after surgery. The success of microsurgery makes the Gamma Knife a less attractive form of primary treatment, particularly of these rather larger tumours. However, the evidence suggests that these tumours are relatively radiosensitive and doses as low as between 20 to 40 Gy have produced control of the endocrinopathy defined as achieving a growth hormone level of less than 5 µg/l. Thus, the Gamma Knife appears to have a useful part to play as an adjunct in the treatment of acromegaly, with the advantage that there is a reasonable chance of also reversing the endocrinopathy. Figure 12.1 illustrates residual tumour, in the cavernous sinus, following trans-sphenoidal microsurgery. The surgically awkward location, relatively far from the visual pathways makes such tumours attractive objects for Gamma Knife surgery.

Nelson's Syndrome

Here again the Gamma Knife has been described as producing a reduction of symptoms in a proportion of patients. In the three patients treated in Bergen a definite reduction in tumour volume has been seen in two and no further growth in the third. Moreover, there has been a reduction in ACTH production in the two which have reduced in size, though normal ACTH values have not been observed. However, arrest of tumour growth has again been the aim in these neoplastically active pituitary adenomas and this seems to have been achieved.

Non-Hormone Producing Pituitary Adenomas

The place of the Gamma Knife in this situation remains to be determined. In general, the tumour will be too large for radiosurgery, since the majority of these patients present with visual symptoms. Moreover, in this situation the chiasm is stretched over the edge of the

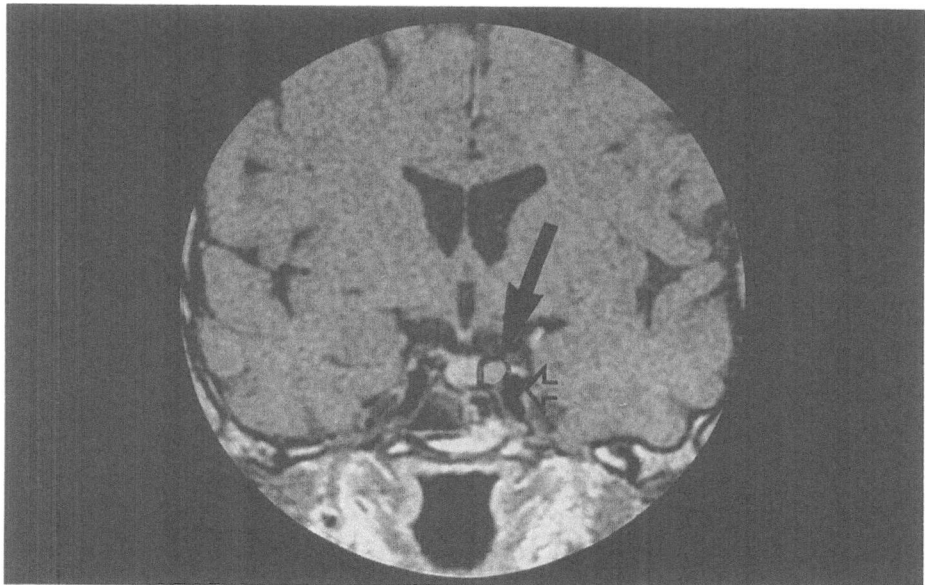

Fig. 12.1
MRI of a patient with acromegaly with a residual *micro*adenoma in the left cavernous sinus. The tumour (arrow) is marked in by an encircling line and the left carotid artery is indicated by an open arrowhead. Thus the intracavernous location can be appreciated. In terms of surgical difficulty, this would be a most awkward tumour for re-operation. Moreover there is much evidence to indicate that re-operation has a much lower chance of controlling acromegaly than a first operation. This tumour is small, and low lying. Thus it is possible to make an appropriate dose plan without exposing the visual pathways to more than 10 Gy. In the author's view, the intracavernous location is a most attractive indication for Gamma Knife surgery in the management of pituitary tumours, for the reasons given above

tumour and is subject to compression. This compressed chiasm tissue will receive the same dose as the tumour edge. This is a quite different radiobiological situation than one in which a normal chiasm is touched over a short length by a radiation dose not exceeding 10 Gy; the usual condition pertaining during Gamma Knife treatment of microadenomas

Nonetheless, while the Gamma Knife may have a part to play as a back-up treatment for a small regrowth following radical surgery (Fig. 12.2), for larger tumours it is probably going to have to share its role with conventional radiotherapy. There are questions involved here which today are impossible to answer. Nonetheless, with the increased

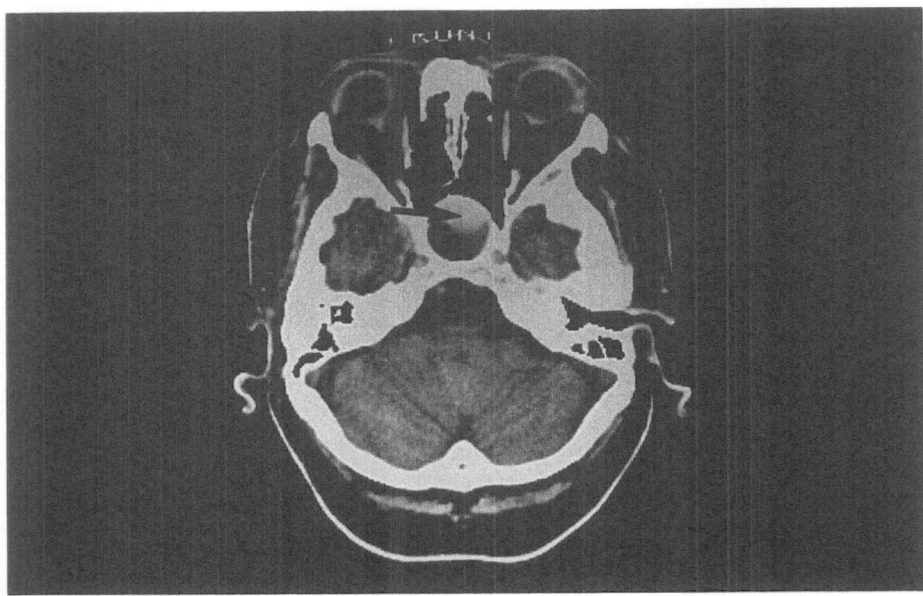

Fig. 12.2
CT of a patient with residual tumour (arrow) following surgery for a
pituitary *macro*adenoma. The low lying location, away from the visual pathways and the easy target definition makes this patient a suitable candidate for
Gamma Knife surgery than an unoperated patient. Thus the Gamma Knife
seems to have a significant role in providing precisely defined radiation for
early detected post-operative residual tumour in cases with an macro-
adenoma

precision granted by modern radiological techniques, follow-up
enables early treatment of residual tumour following surgery, when
the volume is small enough for radiosurgery. This is a promising
approach because conventional radiotherapy is not without risks in
this region. There is a cautionary note that should be sounded at this
point. In some patients, in poor general condition, it may be
considered that surgery may be avoided, with advantage, by using the
Gamma Knife. If this view is accepted, great care must be taken that the
radiological diagnosis is sufficiently secure for treatment to be
initiated on the basis of the radiological findings. If this is not the case,
surgery is still probably preferable. The radiological differentiation
between non-secreting pituitary adenomas, meningiomas in and
around the sella and some craniopharyngiomas is by no means simple.

Craniopharyngiomas

The great majority of these can be treated by intracystic Yttrium-90 installation. Backlund has demonstrated that this gives excellent results in the great majority of patients, with an acceptably low morbidity and mortality. While the author does not expect all his colleagues to agree with him, he is personally convinced, after seeing a number of these patients that the Yttrium-90 treatment is the treatment of choice, for the cystic component of craniopharyngiomas.

However, it is part of Backlund's management concept that the treatment of craniopharyngiomas should be tailored to the patient's needs rather than to the personal preferences of the responsible surgeon. Thus, since a number of craniopharyngiomas are non-cystic, other methods will be necessary for such tumours. In a few, open surgery will be appropriate. In the majority, the solid component will be able to be controlled by radiosurgical techniques. The number of patients treated in this way is small, but the results indicate beyond doubt that craniopharyngiomas are radiosensitive and that Gamma Knife surgery is a most useful treatment, also in this disease.

Suggested Further Reading

1. Backlund E-O (1979) Stereotactic radiosurgery in intracranial tumours and vascular malformations. In: Krayenbühl H, et al (eds) Advances and technical standards in neurosurgery, vol 6. Springer, Wien New York, pp 1–37
2. Backlund E-O, Axelsson B, Bergstrand C-G, Eriksson A-L, Norén G, Ribbesjö, Rähn T, Schnell P-O, Tallstedt L, Sääf M, Thorén M (1989) Treatment of craniopharyngiomas – the stereotactic approach in an ten to twenty three years' perspective. I. Surgical, radiological and ophthalmological aspects. Acta Neurochir (Wien) 99: 11–19
3. Sääf M, Thorén M, Bergstrand C-G, Norén G, Rähn T, Tallstedt L, Backlund E-O (1989) Treatment of craniopharyngiomas – the stereotactic approach in an ten to twenty three years' perspective. I. Psychosocial situation and pituitary function. Acta Neurochir (Wien) 99: 97–103
4. Backlund E-O (1972) Studies on craniopharyngiomas III. Stereotaxic treatment with intracystic yttrium-90. Acta Chir Scand 139: 237–247
5. Backlund E-O (1972) Studies on craniopharyngiomas IV. Stereotaxic treatment with radiosurgery. Acta Chir Scand 139: 344–351
6. Backlund E-O, Rähn T, Sarby B, De Schryver A, Wennerstrand (1972) Closed stereotaxic hypophysectomy by means of Co-60 radiation. Acta Radiol 11: 545–555
7. Burch W (1983) A survey of results with transsphenoidal surgery in Cushing's disease. New Eng J Med 308: 103–104

8. Degerblad M, Rähn T, Bergstrand G, Thorén M (1986) Long-term results of stereotactic radiosurgery to the pituitary gland in Cushing's disease. Acta Endocrinol (Copenh) 112: 310–314

9. Jeffcoate WJ (1988) Treating Cushing's disease. Brit Med J 296: 227–228

10. Nakane T, Kuwayama A, Watanabe M, Takahashi T, Kato T, Ichihara K, Kageyama N (1987) Long term results of transsphenoidal adenectomy in patients with Cushing's disease. Neurosurgery 21: 218–222

11. Ross EJ, Linch DC (1982) Cushing's syndrome – killing disease: Discriminatory value of early signs and symptoms aiding early diagnosis. Lancet II: 646–649

12. Rähn T, Thorén M, Hall K, Backlund E-O (1980) Stereotactic radiosurgery in Cushing's disease acute radiation effects. Surg Neurol 14: 85–92

13. Thorén M, Rähn T, Guo W-Y, Werner S (1991) Stereotactic radiosurgery with cobalt-60 gamma unit in the treatment of growth hormone-producing pituitary tumors. Neurosurgery 29: 663–668

14. Thorén M, Rähn T, Hall K, Backlund E-O (1978) Treatment of pituitary dependent cushing's syndrome with closed stereotactic radiosurgery by means of Co-60 gamma radiation. Acta Endocrinol (Copenh) 88: 7–17

15. Thorén M, Sääf M, Degerblad M, Rähn T, Norén G, Bergstrand CG, Tallstedt L, Backlund E-O (1988) Stereotactic irradiation for pituitary disease. Horm Res 30: 101–104

13. Meningiomas

Introduction

Intracranial meningiomas are one of the commoner intracranial tumours and arise from arachnoid, usually at sites of arachnoid granulations. Thus, the tumours tend to occur in certain defined areas. There is broad agreement that the primary treatment is surgical, not least because these mostly slow growing tumours are usually too big for Gamma Knife surgery, at the time of presentation. Their slow growth permits considerable compensation for their presence, so that they often only give rise to symptoms when they are very large. Meningiomas are peculiar tumours. They do not metastasise outside the skull but they consistently show one of the classic signs of malignancy: they grow across tissue boundaries, to involve dura and bone. Nonetheless, their biological behaviour, despite their pattern of spread is usually benign.

Despite the inappropriateness of many meningiomas for Gamma Knife surgery, the technique still has a part to play. Firstly, success in meningioma surgery is largely related to radical removal. At the time of writing techniques of arterial and venous repair are too inadequate to permit radical removal in many cases. Moreover, meningiomas often arise in elderly patients where the risks of surgery must be weighed carefully against the risks of the untreated disease. With modern imaging techniques, precise observation of tumour growth is possible at relatively small risk to the patient. Thus, despite the desirability of radical surgical removal this aim will be tempered by discretion, both in respect of operating at all and if operation is indicated, in respect of the radicality of removal. It is in this situation that treatment in the Gamma Knife may be appropriate. Moreover, meningiomas have one very attractive characteristic for Gamma Knife therapy: they are easy to visualise and their edge is in general well defined on CT and MRI.

Though the Gamma Knife is increasingly being used to treat meningiomas, its precise role has yet to be defined. The dose requirements for meningiomas in general and for different locations and histological types is at present not determined with any degree of certainty. There is some evidence that meningiomas exhibit a degree of variability of radiosensitivity which tends to hamper optimal dose-planning. These factors together with the slow growth of the untreated tumour means that a number of years must pass before reliable follow up data exist. At present, the type of tumour referred indicates the areas of frustration, where the tumour has proved unresectable.

Types of Meningioma Commonly Referred to the Gamma Knife

Regions where Radical Surgery is Commonly Difficult

This first and foremost applies to meningiomas which have some component in near relation to the cavernous sinus and structures issuing from it. This will include medial third sphenoidal ridge tumours (Fig. 13.1), cavernous sinus meningiomas proper (Fig. 13.2), tuberculum sella tumours (Fig. 13.3), petrous apex tumours (Fig.

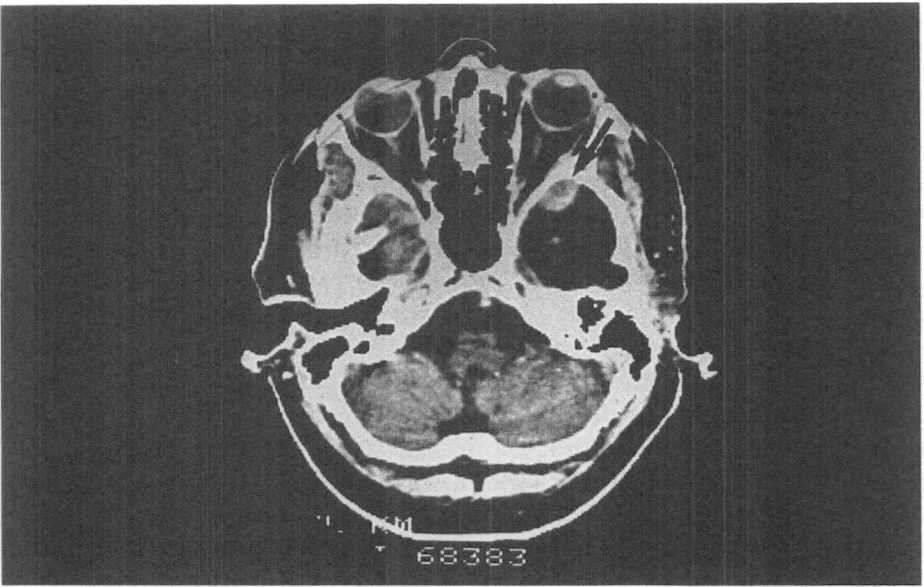

Fig. 13.1
CT of a sphenoidal ridge meningioma (arrow). This patient had been operated and had a postoperative tumour which became visible several years after surgery. He was uninterested in a new operation and it was considered that there was no guarantee that re-operation would be more radical than the first procedure. Thus the Gamma Knife was chosen

13.4). The structures at risk at surgery include the optic nerve, chiasm and tract, the nerves controlling eye movement, the upper brain stem, the cavernous sinus, the carotid artery and its branches and the basilar artery and its branches. The radicality of surgery will depend on the relationship between the tumour and the particular structures at risk with which is in contact. Sometimes a soft tumour peels away easily. Sometimes a firm tumour is so intimately associated with the normal structures as to make radical excision impossible. Even the famous German neurosurgeon, Samii, with his vast experience has suggested that about 20% of meningiomas in contact with the brain stem are incapable of radical removal. Intracavernous meningiomas commonly defy even the most expert surgeons. Moreover, radical surgery, even in the best hands may be associated with an increase in visual defect or diplopia. Another popular site for the Gamma Knife is the tentorium tumour (Fig. 13.5). Other types of meningioma which present problems for radical resection are intraventricular tumours and posterior deep seated falx meningiomas, in the region of the vein of Galen (Fig. 13.6).

Results of Treatment

A number of studies have demonstrated that recurrence of subtotally removed meningiomas is common. Indeed a substantial proportion of seemingly radically removed tumours recur if the follow up period is long enough, for example 15 years. Such a follow up for patients treated in the Gamma Knife does not exist today. The technique is too young. However, the Charlottesville group with a shorter follow up period have demonstrated tumour control, as evidenced by unchanged size or reduction in size in 87% of patients. The Pittsburgh group have shown control over a period of 6 to 36 months in 96% of patients. However, what is interesting in the Pittsburgh material is that 34% of tumours showed reduction in size as opposed to the Charlottesville group, where significant size reduction occurred in only 7% of patients. An examination of our material from Bergen provides some clues in this respect. In this material of 22 patients followed for a year or more, 4 tumours showed a reduction in size and 4 showed an increase in size. All the patients with an increase in size had a dose to the edge of the tumour of 10 Gy or less. All the patients with a reduction in size had an edge dose of 12 Gy. or more. However, low doses at the tumour edge had been the result of either tumour size or of inconveniently placed neighbouring structures. Thus, even when there are risks involved it is important to remember that here as elsewhere, dose is crucial.

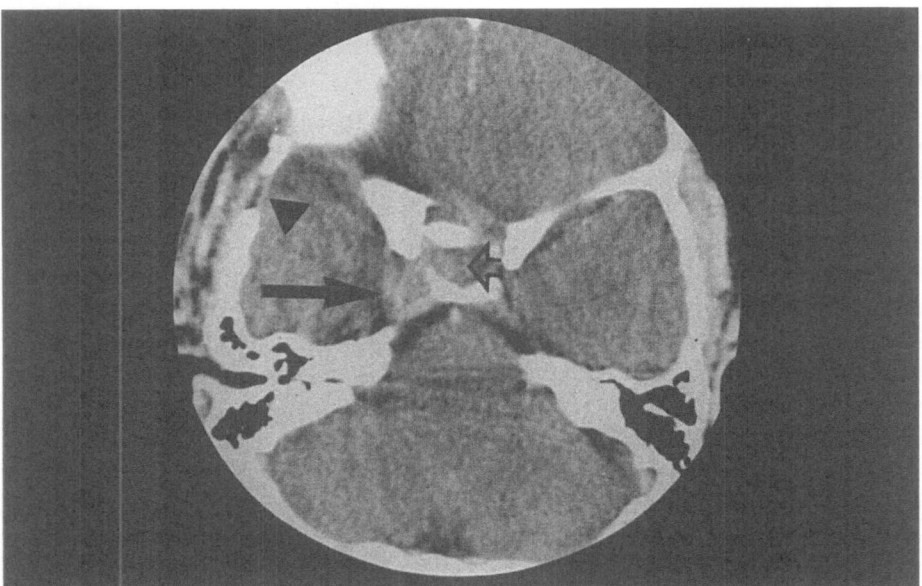

Fig. 13.2. a

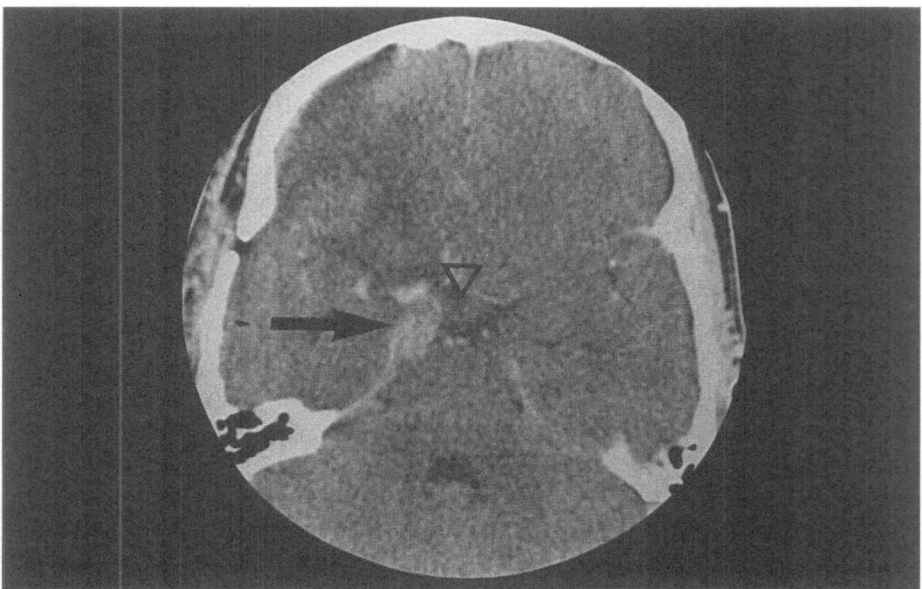

Fig. 13.2. b

Tumours Referred for Primary Treatment

Some cerebello-pontine angle tumours may present when they are relatively small and treatable in the Gamma Knife and are referred for primary treatment. The diagnosis must be certain if they are to accepted without biopsy (see below). Otherwise, high age and poor general condition will result in a small number of olfactory groove and suprasellar meningiomas being referred, where surgery would otherwise be preferable. It should be clear from the preceding section that tumours should only be accepted if an adequate dose can be given. In this context it will be necessary, as indicated above to weigh what is an acceptable risk. It may be, that with some meningiomas, with their tendency to grow and recur, that a slightly higher dose to normal brain may be accepted, than would otherwise be the case.

Tumours which Recur even after Radical Surgery

This applies particularly to a group of tumours which are no longer considered to be meningiomas, the *hemangiopericytomas*. These tumours recur at the site of origin. They also recur at other intracranial sites and can metastasise extracranially. They are moderately malignant and have a poor prognosis. However, palliative treatment in the Gamma Knife can hold them at bay for a while and the treatment can of course be repeated more than once if required.

Fig. 13.2
(a) CT of a cavernous sinus meningioma (arrow). The surgically inaccessible nature of these tumours makes them popular candidates for Gamma Knife surgery. However, it is advisable to reduce the mass as much as possible by open surgery before the Gamma Knife is used. Signs of the previous craniotomy flap may be seen (closed arrow head). In this patient there is spread into the pituitary fossa (open arrow) indicating that the cavernous sinus is not compressed but invaded. **(b)** A second slice from the same CT examination as in Fig. 13.2 a, but 9 mm higher up. The tumour may still be seen (arrow). Moreover, the approximate position of the chiasm is indicated by the open arrowhead. Thus, it is easy to see that the optic tract is at risk. This is an important consideration. If the patient has lost vision on the ipsilateral side, as this patient had, it is particularly important to try to avoid producing a homonymous hemianopsia. In this context not only the chiasm, but in particular the optic tract must be given special consideration during dose planning

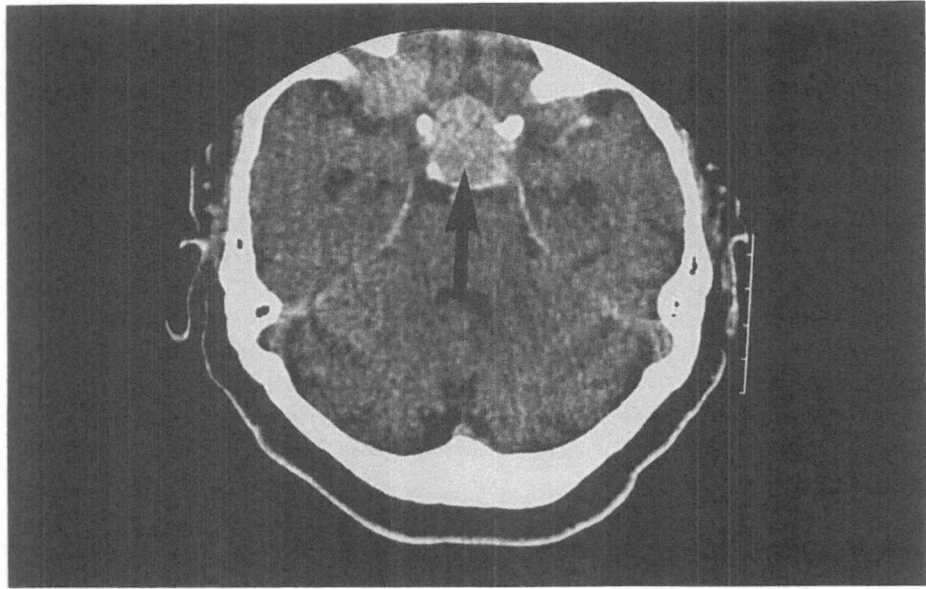

Fig. 13.3
CT of a tuberculum sella meningioma (arrow). This was a very aged patient
where surgery was considered inadvisable. In a younger patient surgery
would have been the first choice

Diagnosis

In most cases the patients will be referred following surgery which was
not as radical as had been hoped. However, in a small number of cases
the referral will be because the patient either is not fit enough for
surgery or refuses surgery. In most cases diagnosis provides no
difficulty. Nonetheless, in a few locations there is a problem related to
the differential diagnosis from other tumours of the region. This is
most likely to be the case in two particular locations. In the cerebello-
pontine angle, differential diagnosis from acoustic neurinoma is not
always straightforward. The other place where difficulties can arise is
with suprasellar meningiomas and their differential diagnosis from
pituitary adenomas. In these two regions great care must be taken to
ensure a reliable pre-treatment diagnosis, by the combination of
physiological and radiological examinations which are currently
available. If it is not possible to reach a decision concerning the
diagnosis after all the tests are performed, it would then be correct to
perform a stereotactic biopsy of the tumour, before treatment is

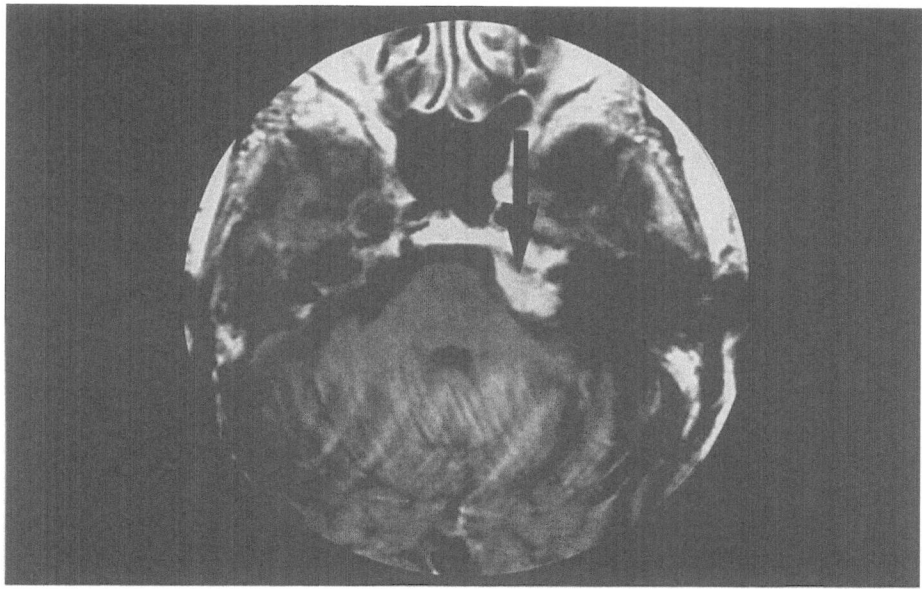

Fig. 13.4
MRI of a petrous apex meningioma (arrow). This conveniently small tumour is a good target for the Gamma Knife. Moreover, this was an aged patient in not very good general condition, so that surgery would have been very risky. In a younger patient, at present, the appropriateness of the Gamma Knife as a primary treatment must be unclear. On the one hand the surgical approach is fairly formidable and the tumours are fairly uncommon. On the other hand, it can be argued that with the intensive radiological follow-up, inherent in the Gamma Knife method, the patient loses nothing by the treatment. Nonetheless, the precise role of the Gamma Knife for petrous apex tumours in younger patients remains to be defined

started. It should be emphasised that this dilemma is going to arise in only a tiny minority of patients but is mentioned because when it arises it is important.

Assessment for Treatment

Tumours Related to the Cavernous Sinus

The aims of treatment are to arrest tumour growth without giving the patient an additional neurological deficit. This requires considerable care in the dose-planning of these tumours, as indicated above. It means that where at all possible, prior debulking of the tumour is al-

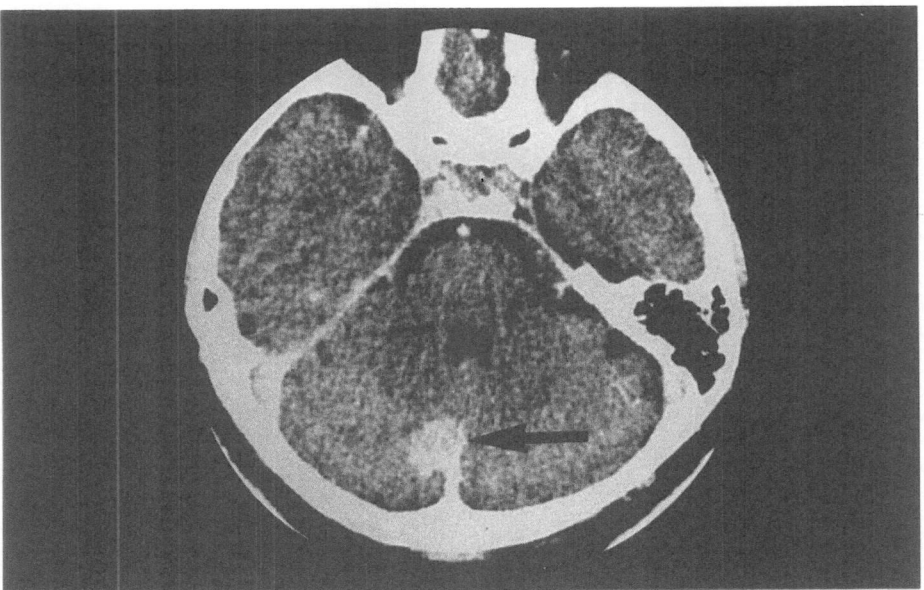

Fig. 13.5
CT of a tentorium meningioma (arrow). This is an excellent object for surgical resection. The author would not have chosen this patient for the Gamma Knife. However, the patient refused to consider any form of open surgery so that the Gamma Knife was the only treatment that could reasonably be offered. It was an extremely simple procedure

most mandatory. Moreover, irrespective of the localizing radiological technique used during the Gamma Knife surgery itself, it is very important to obtain a pre-treatment MRI examination, performed with gadolinium, to determine the full extent of the remaining tumour. This examination is particularly good at revealing tumour spread en plaque, which can be difficult to see on the CT and even at operation. CT is also necessary not least to assess bone thickening, where this is relevant. In some cases an angiogram will be desirable as it can help in the differential diagnosis.

With the anterior tumours, the optic nerve, chiasm and particularly the optic tract are the cause of most concern, and in Bergen it is the practice to avoid giving more than 10 Gy to these structures. This can limit the desired dose to the main body of the tumour, when there is spread of tumour along the optic canal or when it is in contact with the optic tract over a wide area. In such circumstances, extra careful follow-up will be necessary.

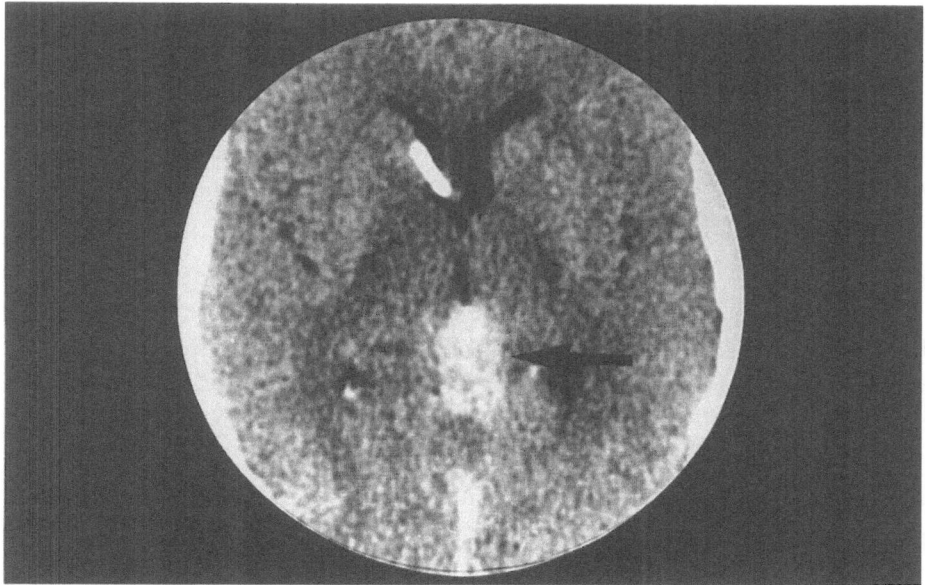

Fig. 13.6
CT of a falx meningioma (arrow). This is a tumour which may be resected but lies in a very awkward location. It also is an excellent target for the Gamma Knife. If the treatment does not work then surgery may be attempted. In this case the patient had already had an recurrence following surgery. It was a most aggressive tumour and the Gamma Knife treatment was not successful. However this does not detract from the fact that such tumours are very attractive targets, always provided that they are not so aggressive as to defeat all forms of therapeutic intervention

With the posterior tumours it is the upper brain stem or even the pons which can be at risk. The dose gradient at the edge of the tumour is less sharp with big lesions, as has been described earlier. It is clearly an advantage with a technique which bases its success on a sharp dose fall at the tumour- normal tissue interface, to have as sharp a dose fall as possible and thus a small lesion. It is this fact that makes pre-treatment debulking desirable, particularly in posterior tumours.

In anterior tumours the purpose of debulking is twofold to reduce tumour volume and also to reduce the irregularity of the tumour shape, so characteristic of these lesions.

While the optic pathways and the brain stem should not be exposed to high radiation dose, to the best of current knowledge the carotid and basilar arteries and nerves controlling ocular movement, at least within the cavernous sinus appear to be resistant to the effects of radiation. Also, while the trigeminal nerve may be damaged by treatment of a cerebello-pontine angle tumour it seems to be relatively resistant in the context of tumours related to the cavernous sinus. It may be mentioned, that it can be difficult to determine whether post-treatment cranial nerve deficit is an expression of radiation damage or continuing tumour growth.

Other Locations

In these circumstances the major factors taken into account at assessment are the size of the tumour with intraventricular cases and in those with hemangiopericytomas the multiplicity of the lesions. The decision to accept a patient for treatment will then be based on the spread of the dose in undesired directions.

Dose

The optimal dose for meningiomas has not yet been discovered. It may not be possible to arrive at a consensus, for this very hetero-geneous groups of tumours. There is some evidence that the dose to the edge of the tumour should not be less than 12 to 15 Gy, but this is at present only a tentative finding. It is obviously desirable to give a higher dose. However, the location and size of the tumour will often make this impossible.

Patient Follow-up

Patients are followed up with CT or MRI every six months for a year and thereafter once a year. It seems sensible at the present time to advise repeat examinations for at least 10 years, because of the known slow growth of these lesions. In the future it may be possible to reduce the frequency or the duration of these follow up examinations, but such a change will be predicated on more precise information on the results of treatment for these tumours than is available today.

Suggested Further Reading

1. Kondziolka D, Lunsford LD, Coffey RJ, Flickinger JC (1991) Stereotactic radiosurgery of meningiomas. J Neurosurg 74: 552–559
2. Steiner L, Lindquist C, Steiner M (1991) Meningiomas and Gamma Knife radiosurgery. In: Al-Mefty O (ed) Meningiomas. Raven Press, New York, pp 263–272

14. Miscellaneous Indications for Gamma Knife Radiosurgery

Introduction

This chapter concerns cerebral metastases, together with a group of different diagnoses which individually either occur seldom or are referred seldom, so that general principles have not as yet been worked out in respect of them. Nonetheless, many of these diagnoses have characteristics which require that the referring physician is in possession of specific information, in order to assess the relevancy of referral and to be in a position to communicate with the patient.

Malignant Tumours – Metastases

Radiation treatment of intracranial malignancies may be either primary or may follow surgery as an adjunct. If radiosurgery is to be used then there are three types of radiation treatments currently available. The size of these lesions may preclude Gamma Knife surgery. Nonetheless, the majority are smaller tumours, making it a relevant alternative in most cases. It is conventionally taught that cerebral metastases visible at CT or MRI examination are usually associated with other as yet invisible metastases. This teaching is used as an argument for giving fractionated whole brain radiation, which is in part therapeutic and in part prophylactic. However, there are disadvantages to this technique. Firstly, in one large study, the average survival following radiotherapy alone for cerebral metastasis was about 20 weeks. The survival following surgery and whole brain radiotherapy was about 40 weeks. However, the local recurrence rate was 20%. Moreover, an appreciable portion of the remaining time available to the patient has to be spent in hospitals. Secondly, once whole brain irradiation has been given, subsequent conventional radiation, with adequate doses will be compromised. At the present time there are two main methods for Gamma Knife treatment of metastases. In published results from Pittsburgh it has been used as a booster, following conventional radiotherapy. The Stockholm group on the other hand has used the Gamma Knife as the primary treatment for cerebral metastases.

The advantages of the Gamma Knife are obvious, provided that its clinical effect is adequate. Firstly, the patient only has to stay a maximum of forty eight hours in hospital. Secondly, subsequent metastases can be treated, despite the earlier radiation, because the dose volumes are so limited and specific. The results outlined below are obtained from the Karolinska Hospital. There have been one to three metastases present at the first treatment and in a number of patients subsequent treatments have been necessary. There have been problems in a few patients with low dose, because of a brain stem location or because of subsequent haemorrhage in the tumour. However, there have been hardly any cases of death that could, with certainty be attributed to the cerebral metastasis(es). Moreover, there has been almost no local recurrence. Thus, cerebral metastases would seem to be a most promising field for Gamma Knife surgery, bearing in mind the limitations inherent in the systemic disease.

The commonest brain metastases come from carcinomas of the bronchus, breast and kidney. Carcinoma of the thyroid and melanomas also occur sometimes. In addition, there are a variety of other primary tumours which occasionally metastasise to the brain. It has been shown that good results can be obtained with all these categories.

Clinical Presentation

The clinical presentation will consist of the symptoms of raised intracranial pressure, headache, vomiting and depressed level of consciousness, together with focal neurological deficits dependent on the location of the tumour.

Aims of Treatment

The primary aim of treatment is to stop tumour growth. A reduction in volume or disappearance of the tumour are of course even better but are not a required indicator of successful treatment, as it is performed today.

Patient Assessment for Gamma Knife Surgery

The diagnosis is easily made in patients with a known cancer of the right sort. However, in a substantial proportion of cases, the metastasis will be the first presentation of the tumour. In these circumstances a biopsy is to be preferred prior to treatment, if the primary tumour cannot be found on routine simple clinical screening. The clinical screening should include a chest X-ray, a urine analysis, an examination of the breasts, a rectal examination and an ultrasound examination of the abdomen. The chances of obtaining a positive diagnosis if these tests are negative appears to obey a law of diminishing returns. It has been the practice in Stockholm to exclude from treatment those patients who have a life expectancy from their systemic disease, of less than two months.

Dose

While the ideal dose is not yet known, the aim is to keep the edge dose above 25–30 Gy. Some tumours will, necessarily, be complicated by post-irradiation oedema, at these dose levels. This is considered acceptable. It is contended that it is better to have a little oedema, rather than to fail to control the tumour. Such oedema is seldom a practical problem and is managed by dexamethasone, on an ad hoc basis.

Acceptance of Patient for Treatment

The patient with known cancer, either discovered previously or at routine screening, and a metastasis that is preferably not more than 30 mm in diameter is the ideal case for treatment. The anatomical location is only a difficulty for superficial lesions, near the frontal or occipital poles and even these can be managed with correct placement of the stereotactic frame on the head. If the diagnosis is uncertain and routine screening has given no result, then another strategy is necessary. If Gamma Knife surgery has been decided upon then a stereotactic biopsy is performed with subsequent Gamma Knife treatment, at the same session, if the diagnosis of metastasis is confirmed.

Malignant Tumours – Gliomas

Type of Tumour

In this context, the relevant diagnoses are glioblastoma and anaplastic astrocytoma. Here a biopsy is mandatory. Moreover, these are not easy objects for the Gamma Knife because they most often do not have a well defined edge either radiologically or in reality. In general, they are larger than desirable for this form of treatment. However, the results of conventional treatment are so disappointing that any new form of management needs to be examined. At present, there is broad agreement that the Gamma Knife should only be used in association with conventional, fractionated radiotherapy in these cases. In other words it has a booster function.

Clinical Presentation

The clinical presentation will consist of the symptoms of raised intracranial pressure, headache, vomiting and depressed level of consciousness, together with focal neurological deficits dependent on the location of the tumour.

Aims of Treatment

The primary aim of treatment is to stop tumour growth. A reduction in volume or disappearance of the tumour are of course even better but is not considered a necessary indicator of successful treatment, as it is performed today.

Results of Gamma Knife Surgery

This is unknown today. Early experience from Pittsburgh, presented at the 2nd International Radiosurgery Symposium, consisted of 10 glioblastomas and 5 anaplastic astrocytomas. The Pittsburgh group were cautiously optimistic. However, it is still too early to assess the results and compare them with other forms of treatment, including combinations of radiotherapy with surgery and with chemotherapy.

Patient Assessment for Gamma Knife Surgery

The requirement is a tumour which is reasonably easy to localize on CT or MRI and which is not more than 30 to 40 mm maximum diameter. This can also apply to remains of glioblastomas left after surgery. It can also apply to tumours diagnosed while still small, because of a deep location which results in early symptom production and thus early clinical presentation. Such a tumour may well not be amenable to surgery. The largish volume of most malignant gliomas accepted for treatment is justifiable, because with the very poor prognosis a higher treatment related morbidity can be tolerated. At present, the role of the Gamma Knife in these cases remains undetermined. Further documentation of the effects of treatment is required, before the treatment can be recommended for other than specially selected cases.

Follow-up

For these patients, the follow-up radiological examinations must be taken every three months for obvious reasons. It may be very difficult to distinguish between radionecrosis and changes due to the tumour, with the technology at present available. Thus, it is very important to correlate X-ray changes with clinical changes, which should be recorded at the same time as the X-ray studies are taken.

Malignant Tumours – Optic Melanoma

Clinical Presentation

These are detected by the ophthalmologist and confirmed with computer imaging studies. While the amount of data is at present small, they would appear to be excellent objects for Gamma Knife radiosurgery.

Aims of Treatment

The aim of treatment is to remove the tumour so that it is invisible both radiologically and at ophthalmoscopy, while preserving useful vision. Clearly, there will be a scotoma due to the presence of the tumour itself. While this scotoma will remain unchanged, it should ideally not increase in size. Even so, the dangers of this condition are so profound, that while retention of vision is both highly desirable and also attainable, it is not a sine qua non. It has to be better to aim for control of the tumour and risk visual damage, than to be over careful in order to ensure that vision is retained.

Patient Assessment for Gamma Knife Surgery

The conventional treatment is excision of the eye. However, radio-active ruthenium implants have also been used in an attempt to reduce the trauma resulting from exenteration of an orbit, with its disturbing cosmetic consequences. The trouble with ruthenium is that, as a beta emitter, the therapeutic radiation dose will not extend far beyond the base of the tumour. This means that it can be difficult to achieve tumour control without producing retinal damage. The Gamma Knife is capable of causing optic melanomas to disappear, without further loss of vision. However, the role of radiosurgery in preventing metastasis and prolonging life remains to be determined. It may be that it should be used in association with a beta emitter. The patient will need to be warned that the eye will be fixed under local anaesthetic prior to treatment.

Acceptance of Patient for Treatment

There is one overriding factor that must be assessed when deciding if a patient is suitable for treatment. This is the size of the head. Measurements must be made using *CT or MRI examinations of the whole head* at the level of the orbit. Thus, such examinations are mandatory if a patient is to be accepted. Some patients with large heads will not be appropriate, because the large head, with this extremely eccentrically placed lesion can, very rarely defeat attempts to place the lesion satisfactorily within the Gamma Knife helmet. Even so, with some ingenuity in placement of the frame, nearly all patients can be treated for this condition.

Follow-up

This is performed by the referring ophthalmologist as often as is deemed necessary. The ophthalmological progress or regress of the tumour is noted together with signs of radiation damage to the retina. In addition, the visual fields and visual acuity are recorded. This is vital as the whole point of the treatment is that it potentially allows retention of a **functioning eye.** It is usual to request MRI or CT studies every six months after treatment.

Other Benign Tumours

Gliomas

These are in the main inoperable small **astrocytomas.** They should be confirmed by biopsy. Dose-planning can be really very difficult because of the uncertainty as to the edge of the tumour. Moreover, the dose must not be too high, since there may be important functioning cerebral tissue within what is judged at X-ray to be tumour.

Another constraint on dose is the location of these tumours. The astrocytomas referred to the Gamma Knife in Bergen have been mostly located near to or in the hypothalamus in children. Clearly, with the very slow growth rate the indication for treatment is not clear cut. In general it is desirable to have tumour growth demonstrated prior to treatment. These tumours can be very indolent and since radiosurgery can damage nervous tissue, in cases with a central location the results of brain damage can be catastrophic. However, having mentioned the theoretical problems, the early results in Bergen, for this group of patients would seem to be rather encouraging. Early tumour shrinkage is by no means unusual.

In addition to the astrocytomas individual **ependymomas, oligodendrogliomas** and **gangliogliomas** may be referred. All must be considered seriously and may be suitable for Gamma Knife radio-surgery.

Finally, there are two related types of glioma, which are occasionally referred. These are the **optic nerve glioma** and the **chiasm glioma.** The difficulties related to the chiasm glioma are similar to those of the hypothalamus glioma, as outlined in the previous paragraph. The optic nerve glioma is usually referred in the hope that they may be treated without cosmetic injury and even retention of vision. The advice that must be given is that vision will almost certainly be lost, following a tumouricidal radiation dose. This is not the tragedy it might at first appear, because the untreated disease will undoubtedly destroy vision in the affected eye, as will effective surgery in resectable cases. It is not clear today if the Gamma Knife is a viable alternative to surgery in these rare tumours. Even so, with the precision of modern computer imaging it is not unreasonable to try radiosurgery even in cases which are technically resectable. In cases where the glioma has passed beyond the optic canal – which will in consequence be widened – to the chiasm, Gamma Knife radiosurgery becomes indeed an attractive alternative, with the possibility of halting further growth, without destroying the vision of the contralateral eye. However, dose-planning will require particular care, not only with regard to the chiasm, but also with regard to the ipsilateral optic tract. Nonetheless, as stated above, the rarity of the tumour will ensure that the precise role of radiosurgery will remain to be defined for several years to come.

Non-gliomatous Benign Tumours

These include hemangioblastoma, trigeminal neurinomas and glomus jugulare tumours. Hemangioblastomas are radiologically well defined and provide no special problems in themselves. However, if located in the brain stem, some caution in respect to the dose will be required. It is considered inappropriate to advise radiosurgery for those haemangioblastomas which are associated with a large cyst. The trigeminal neurinoma is treated as any other benign tumour. However, excellent CT pictures of the base of the skull will be required to ensure that the relation of the tumour to the foramen ovale is defined. This is a tumour which may require a particularly low placement of the frame The glomus jugulare tumour will also require a low placement of the frame. Moreover, it is difficult to decide the timing of treatment for these indolent tumours, which may remain unchanged over a very long duration.

Functional Radiosurgery

Parkinsonism

Parkinsonism, which was one of the main reasons for constructing the first Gamma Unit has lost a lot of its popularity. Firstly, the advent of L-dopa has changed the place of thalamotomy radically. Secondly, the Gamma Knife is a technique which precludes physiological testing. Moreover, design of appropriate lesions is complex with the collimators currently in use and the treatment, is very time-consuming. However, radiosurgery remains a conceptually attractive alternative with its closed technique. It will be interesting to see if the data acquired during open stereotactic operation with physiological testing will permit the placement of lesions using anatomical landmarks alone. The anatomical resolution of a modern MRI machine makes this a credible possibility. There are some most recent studies, not yet in print which indicate that Parkinsonism may well be on the way back as an indication for Gamma Knife surgery.

Trigeminal Neuralgia

Trigeminal neuralgia was treated by Leksell in the early days of Gamma Knife surgery. However, while initially promising, the results proved impossible to assess, because of a large number or technical problems. More recently interest has been revived for this indication. There are some early results, not yet in print which would seem to suggest that this is yet another field where the Gamma Knife will have a part to play in the future.

Cancer Pain

Here the indication would appear to be more definite. Not only may the Gamma Knife be used for a pituitary anterior lobe ablation but it may also be used for thalamotomy for unilateral pain. The patients requiring destructive lesions of the anterior hypophysis or the thalamus are desperate and have a short life expectancy. This makes the stress and risks and hospital stay, necessitated by open surgery particularly undesirable. Moreover, in respect of thalamotomies, it is more acceptable, for the same reasons, to accept the risks inherent on treating without the benefit of pre-operative physiological testing. Because the life expectancy is short, the secondary central pain, arising one to two years after destruction of central nervous tissue will not be a problem. This is a treatment which can be recommended unreservedly so long as its limitations are born in mind. These are that it is not uniformly successful and it is only to be used for unilateral malignant pain. However, with experience from a larger number of cases, the number of successes could be expected to improve. This is the usual pattern observed in destructive lesions for pain. With experience the success rate rises.

Psychosurgery

Psychosurgical intervention has been somewhat under a cloud in recent years. This is partly due to the development of improved pharmacological agents and partly due to changing ethical considerations. It is considered that focal brain destruction for psychological symptoms is an unappealing therapeutic technique. It is also felt, by many, that this is interfering with the creation. The author thinks that it is self evident that psycho-surgery has to be a treatment of last resort. It is the management of desperation. Thus, the number of patients who will qualify is very limited. Nonetheless, a few tortured souls will qualify and it is then sad if a useful treatment is denied, when all else has failed.

In this situation, a Gamma Knife lesion represents an elegant alternative, providing that it safe and reliable in relieving symptoms. Safety is of course one of the hallmarks of Gamma Knife surgery. With regard to reliability, a recent study by Mindus et al. (1987) showed that it could be effective treatment, for management resistant anxiety syndromes. In this paper the mean duration of symptoms was 17 years (range 5 to 25). This would appear to be reasonable evidence that these patients were genuinely treatment resistant. The radiosurgery was satisfactory in those patients where a lesion in the anterior internal capsule could be demonstrated, on both sides. However, the dose required to produce the desired brain lesion without producing unwanted cerebral damage outside the target lesion volume, is not yet settled. As a result the dose in the above-mentioned series was on the careful side. In consequence, two of the seven patients no lesion on MRI and they also did not experience symptom relief. It must be emphasised that failure was due to an inadequate effect of radiosurgery not to an adequate lesion production without clinical effect. On the contrary, when the desired lesions were produced, the patients were consistently relieved.

It is the author's contention that Gamma Knife radiosurgery for functional disorders is the most under used aspect of the technique at the present time. It is to be hoped that an increasing number of desperate people will be offered relief from their suffering in the future. With the Gamma Knife, as far as we know, these patients risk only minor discomfort for a few hours. However, it will remain mandatory that the choice of patients remains rigourous, if this aspect of Gamma Knife treatment is not to be brought into disrepute.

Suggested Further Reading

1. Backlund E-O (1979) Stereotactic radiosurgery in intracranial tumours and vascular Malformations. In: Krayenbühl H, et al (eds) Advances and technical standards in neurosurgery, vol 6. Springer, Wien New York, pp 1–37
2. Kihlström, L, Karlsson B, Lindquist C (1992) Gamma Knife surgery in brain metastases. Proc. International Symposium on Radiosurgery (Pittsburgh). Elsevier, New York, pp 429–434
3. Leksell L, Larsson B, Andersson B, Rexed B, Sourander P, Mair W (1960) Lesions in the depth of the brain produced by a beam of high energy protons. Acta Radiol 54: 251–264
4. Lindquist C, Hindmarsh T, Kihlström L, Mindus P, Steiner L (1992) MRI and CT studies of radionecrosis development in the normal human brain. In: Steiner L (ed) Radiosurgery: baseline and trends. Raven Press, New York, pp 245–253
5. Lindquist C, Steiner L, Hindmarsh (1992) Gamma Knife thalamotomy for tremor: report of two cases. In: Steiner L (ed) Radiosurgery: baseline and trends. Raven Press, New York, pp 237–243
6. Mindus P, Bergström K, Levander SE, Norén G, Hindmarsh T, Thuomas K-Å (1987) Magnetic resonance images related to clinical outcome after psychosurgical intervention in sever anxiety disorder. J Neurol Neurosurg Psychiatry 50: 1288 – 1293
7. Moser RP, Backlund E-O (1984) Stereotactic techniques in the treatment of pineal region tumors. In: Neuwelt EA (ed) Diagnosis and treatment of pineal region tumours. Williams and Wilkins, Baltimore, pp 236–253

15. The Future

Introduction

The main aim of this final short chapter is to ask some questions and indicate some possibilities. It is necessarily speculative, but since the Gamma Knife has now treated nearly 11,000 patients and is established in three of the five continents, it is relevant to ask "Quo vadis"? The machine was introduced at the end of the 1960s to treat functional disorders of the brain and is used now, for the most part, to treat tumours and malformations. It was initially designed to treat very small volumes. It is unlikely that it will be able to treat larger targets in the future than those that are being attacked today. This is because, as has been mentioned repeatedly, there is a biological limit to the size of lesion appropriate to attack with radiation, because of the increasing risk to normal tissue with increasing target volume.

It can be safely stated that nobody involved with radiosurgery in the early days could have foreseen where it would have come. By the same token it is not easy to be sure of the place of the Gamma Knife in fifteen or twenty years time. Nevertheless, studies are being carried out at a variety of different centres: studies that will contribute to shaping the future of Gamma Knife surgery. In the sections that follow some of the areas in which new knowledge is being sought will be described in outline.

Target Limitations

In general terms a Gamma Knife target is limited as to size, because of the increasing risk to normal brain tissue, with increasing target volume. In some few cases a peripheral target location may also provide some difficulties. Attempts to improve the design of the instruments used in Gamma Knife surgery, in order to reduce target related difficulties are the subject of current attention by the manufacturers, Elekta Instrument AB in Sweden. However, other measures may be taken to optimize the use of the current equipment in regard of gaining access to inaccessible targets. Amongst these, an optimal design of the treatment suite is important. In respect of optimal frame application, it is an advantage to have the frame application room next to the gamma unit. If the application and treatment rooms are next to each other it is a simple matter to check frame placement prior to radiological examination. If necessary the frame can be adjusted at this stage.

The Reemergence of Functional Indications for Gamma Knife Radiosurgery

The existing functional indications are as relevant as they have ever been, though probably they are not as well known as they might be. Moreover, Professor Lindquist, at the Karolinska Hospital in Stockholm is exploring new aspects of functional treatment. He has been using the Gamma Knife in the treatment of epilepsy in association with the use of magnetoencephalography. This raises all sorts of interesting new perspectives. It seems reasonable to hope that neurologists, in particular will like the idea of closed treatment for epilepsy severe enough to require surgery. Open operation, for intractable epilepsy, in correctly selected cases is a very demanding long-lasting procedure. In addition Professor Robert Rand at the Good Samaritan Hospital in Los Angeles, California is also identifying a greater role for the Gamma Knife in functional disorders, such as Parkinsonism and Tic Douloureux.

Quantification of the Effects of Treatment

This is an area which requires increased precision and the group in Pittsburgh has attacked this problem from their first publication. This group has produced an impressive number of high quality publications on two main areas. Firstly, their clinical reporting has been detailed and characterized by the enumeration of a large number of parameters. Moreover, there is an internal consistency between their publications, so that these papers set a standard and greatly facilitate comparison for those of us who also wish to publish results

The other quantification, that is being studied in Pittsburgh relates to an attempt to quantify the prediction of the risk of brain damage, following Gamma Knife surgery. To do this they have used mathematical models and observed the rate of complications, following the treatment of arteriovenous malformations. This is of course an attempt to quantify the effects of the brain integral dose. They point out that when assessing the dose to the brain, dose histograms are more useful than simple measurements of dose at the lesion edge. The significance of the Pittsburgh approach remains to be seen but it seems altogether most promising.

Radiobiological Variables

This is an area of research interest in Bergen. Questions relating to the correct dose and the effects of dose rate, dose volume and dose homogeneity remain to be answered. There are many unknown factors relating to the interaction of different shots on each other. However, what is more important than the technical detail of a given research programme is the understanding that better treatment is to be expected, from more efficient delivery of the radiation dose. It is our hope to promote the construction of better dose-plans in the future, as the result of our labours.

Appendix

Introduction

The appendix applies to the changes that have occurred in radiosurgery since 1993. Instead of rewriting the chapters, these changes are described where relevant in this appendix. The reader may gain a sense of the development in the field by reading first the chapter and then the appendix. Those who wish to delve deeper may refer to the extensive reference list (see pp. 171–198). It is organised according to subject.

Chapter 1

No changes.

Chapter 2

The importance of the right angle between the centre beam and the frame and film is emphasized in this chapter (see p. 16 and Fig. 2.9). Today this is less important. For open stereotaxy and radiosurgery using the KULA system the requirement is removed because of a new device called the Leksell Localiser. This uses a somewhat different fiducial system. If the newer dose-planning system GammaPlan is used, this also compensates for non-rectilinear geometry in the same way as the localiser. The following figures indicate the way in which the newer fiducials work.

The dose-planning system now contains an algorithm which will correct for some degree of non-rectilinear geometry. This is a change which permits simplification of the technique.

The fiducials closest to the X ray tube will be smaller and those furthest away will be smaller as indicated in Chapter 2, Fig. 2.9 (p. 16).

The technique advances all the time.

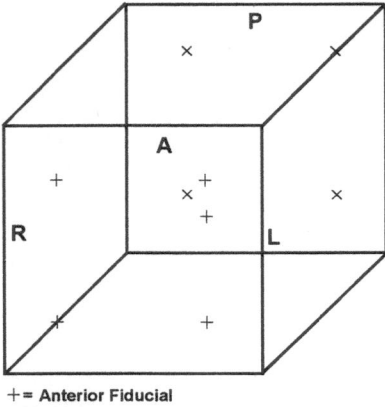

+ = Anterior Fiducial

× = Posterior Fiducial

Fig. a. The position and shape of the fiducials in the angiogram indicator adapted for the newer technology

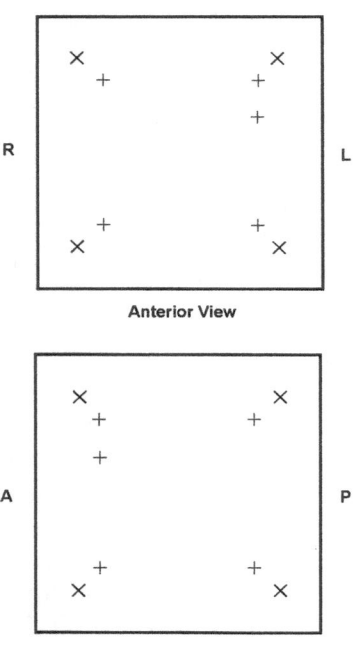

Fig. b. Anterior and lateral view of fiducials. Note the small fiducials are symmetrically placed within the larger fiducials. This represents a central beam, placed at the centre of the fiducial system and at a right angle to it. It is ideal but not essential. The X-ray tube is at the back in the AP film and to the right in the lateral film. Note the extra fiducial at the front identifying the left anterior position

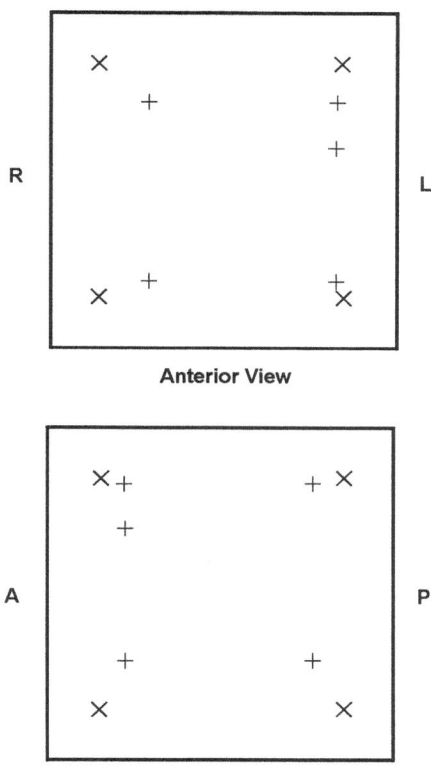

Anterior View

Lateral View

Fig. c. AP and lateral view of fiducials. The X-ray tube has the same positions as in Fig. b. In this diagram the smaller fiducials (proximal to the X ray tube) are within the larger fiducials but not symmetrical. This is acceptable and the dose planning algorithms will accommodate for these asymmetries

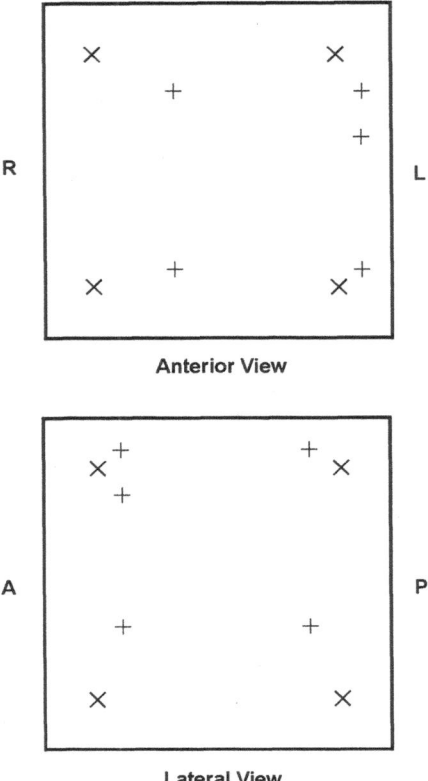

Anterior View

Lateral View

Fig. d. Anterior and lateral view of fiducials. The X-ray tube is at the back in the AP film and on the right in the lateral film. Here the smaller fiducials are outside the larger. While the software *may* accommodate for this position, it is undesirable, avoidable and should not be permitted

Chapters 3 – 5

No changes.

Chapter 6

At the end of this chapter mention is made about the concerns of geometrical inaccuracy in MRI images. While it remains true that all attempts should be made to keep the target central within the stereotactic space, the concerns about MRI accuracy have lessened with time.

At present head coils have replaced body coils for dose-planning MRI images. The errors to be expected and the ways to counteract them are now reasonably well documented. Moreover, the newest dose planning software will simply not accept inaccurate images, thus compelling users to calibrate their imaging equipment to produce geometrically adequate images.

The concerns about MRI inaccuracy remain amongst a number of users. This may in part be related to the inaccessibility of modern MRI equipment at specific locations. It is to be hoped that this situation will change with time. The accuracy of the MRI does depend on the use of excellent modern equipment and those who perform radiosurgery should be aware of this. It is also important to remember that the treatment cannot be more accurate than the images on which it is based and that all kinds of inaccuracies associated with the method are additive and not exclusive so that every inaccuracy must be avoided since it will contribute to the total inaccuracy.

Chapter 7

Fractionation and Radiosurgery

There is no alteration to make here. However, in the last three years there has been an extensive and as yet unresolved debate concerning the role of fractionation with focussed radiation as an addition to single session treatment. This approach is mainly advocated by oncologists using linear accelerators. However, Inoue et al. (see References) have applied hypo-fractionation to the treatment of gliomas, using the gamma knife.

The author does not know what the place of fractionation will be in the future. It does seem that it is most relevant in the palliation of glial tumours and thus is not quite in the main-stream of the gamma knife method. It has been pointed out that the results of gamma knife radiosurgery are so good using single session treatment, that it would be very difficult even with a huge number of patients to determine any advantages of fractionation.

Linear accelerator and particle accelerator users have employed hypo-fractionation techniques for a long time. Dr. Steinberg from California has long treated pituitary adenomas with a Helium ions, using a cross-over beam technique and 4 or 5 fractions. Dr. Colombo from Vincenze in Italy has divided the dose into more than one fraction in the treatment of a number of AVMs. The work of such authors does not clearly indicate an advantage for hypo-fractionation nor does it indicate any disadvantage, other than the increased time and complexity of the treatment.

Thus, readers should be aware that this question is discussed repeatedly without any consensus having been reached as yet.

Dose Homogeneity

Another topic which was the subject of much attention is the importance of the homogeneity of the dose of radiation. Three years ago this was a source of great concern to radiation therapists. As the field of radiosurgery has matured and more radiation therapists have become involved in it, the significance of homogeneity appears to have decreased. It is pointed out that the regions of non-homogeneous dose all occur within the target volume. Moreover, it is realised that lack of homogeneity is an inevitable price which must be paid with current technologies to achieve a dose which conforms to the shape of the target.

Chapters 8 and 9

No changes.

Chapter 10

Arteriovenous Malformations

The End-Point of Treatment for AVMs

There has been much discussion over the last two years about this topic. It has been suggested repeatedly that it may not be necessary to obliterate the entire AVM to obtain protection from further bleeding. In particular it has been suggested that the presence of just an early filling vein may represent a satisfactory result. However, in 1996 the Pittsburgh group published a paper based on an extensive material which indicated that total obliteration is still the required end-point and that anything less is unsatisfactory. In this context the obliteration can still be judged safely only by angiogram. It is to be hoped that in the future MRA may develop to the point that MRA technology is widely available at a quality which obviates the need for angiography. However, at the time of writing this stage has not yet been reached.

Angiograms and Dose-Planning

Today, it is advised that dose-planning be carried out with both angiogram and MRI. The reasons for the angiogram are to ensure that the target volume is as small as possible. The reason for the MRI is to ensure the correct location of the dose-volume and to facilitate the adaptation of a dose plan to respect important structures which should not receive high doses of radiation. It is still considered that angiography is vital for adequate dose-planning, however inconvenient this may be.

One major change is that it is now possible to substitute conventional angiography with DSA which enables a much better definition of the malformation. Straightening algorithms for DSA now exist which make this possible.

Radiobiological Details

MRI changes: According to the work of Drs. Yamamoto and Tanaka from Tokyo, it would seem that malformations which have been shown to be obliterated on angiography continue to change on MRI. Thus, obliteration may be a clinical end-point but the obliterated malformation continues to change for years afterwards. The significance of this is at present not known.

Normal Arteries: One patient came to post-mortem and it was shown that normal arteries outside the malformation showed radiation damage, even though they were not occluded.

Age of Patient: Comparison between children and adults indicated that radiosurgery in children gave a higher success rate with a lower complication rate than in adults.

Size of Malformation: The Tokyo group mentioned above and the Mayo clinic group have both looked at AVMs treated with an almost identical regimen as indicated in the table below.

Yamamoto et al.		
Volume	Dose (Gy)	Obliteration rate%
$\leq 4.1\ cm^3$	20	74.5
$\geq 4.2\ cm^3 - 14.2\ cm^3$	18	74.4
$> 14.2\ cm^3$	16	75.0

Coffey et al.		
Volume	Dose (Gy)	Obliteration rate%
$\leqq 4.1$ cm^3	20	Overall – 72.1
$\geqq 4.2$ cm^3 – 14.2 cm^3	18	
> 14.2 cm^3	16	$\geqq 10$ cm^3 – 71.4

These figures indicate that larger malformations may be treated with an acceptable result. The rather strange choice of volumes for the different categories relates to the earlier practice of recording size in terms of lesion diameter. 4.1 cm^3 is the volume of a sphere of 2 cm diameter and 14.2 cm^3 is the volume of a sphere of 3 cm diameter.

The success rate in these two series is rather lower than that usually quoted. This is probably because the prescription dose is on the low side.

Risk Factors: The accumulated experience has led to some reclassification of malformations. Thus Inoue et al. have shown that AVMs may be classified according to their angiographic appearance into three groups, Moya Moya type (diffuse), arteriovenous type and mixed type. The Moya Moya type seems to give the best response.

Pollock et al. have determined the following risk factors in their material. Multivariate analysis revealed three factors associated with a risk of haemorrhage. These were prior haemorrhage, a single draining vein and a diffuse morphology. On the basis of this classification they defined different four groups of risk for haemorrhage. They found an annual rate of bleeding of 0.99% for low-risk AVMs, 2.22% for intermediate-low-risk AVMs, 3.72% for intermediate-high-risk AVMs, and 8.94% for high-risk AVMs.

The above outline indicates that the field of radiosurgery for arteriovenous malformations is advancing and developing.

Cavernous Haemangiomas

This is the current Index Medicus approved term for the lesion which goes under various names, including cavernoma, cavernous angioma, cavernous venous malformation (CVM). It is a major component of the group of Angiographically Occult Malformations or AOVMs.

These lesions remain problematical and are the source of much debate amongst experts. They have been treated systematically by the Pittsburgh group which has applied strict criteria and tended to treat small lesions. Even so, the experience this group publishes is in keeping with that of other groups. These lesions usually receive a relatively low dose (15 Gy to the edge) and those treated are usually 15 mm in diameter or less. Even so there is an incidence of radiation induced complications of as much as 20%.

There is really no other lesion where such a low dose to such a small volume gives such a high rate of complications. For this reason great caution should be exercised in the use of radiosurgery for this indication. In particular this caution should be applied to lesions in the brain stem. Dr. Awai from Yale suggested at the AANS meeting in San Diego in 1995 that the bleeding risk from the small lesions deemed appropriate for radiosurgery was lower than that for larger lesions. This makes the complication risk even more serious.

Chapter 11

Acoustic Schwannoma

There was some debate three years ago concerning the correct dose to these lesions. However, this has now largely been resolved. At a meeting in Lanai in November 1995 there was a discussion between those who wanted to give a high dose and those who favoured a low dose. The difference between these doses was only 2 Gy. Most people today favour a highly conformal dose-plan with a prescription dose between 12 to 15 Gy.

The latest figures from Dr. Norén and others, using current dose-schedules and dose-planning techniques indicate a tumour control rate of 95% and an incidence of temporary facial deficit between 5 and 10%. The higher figures, reported only a few years ago are increasingly a matter of only historical interest.

Chapter 12

Pituitary Adenomas

There is still some debate about the role of radiosurgery for these lesions. However, many centres have had difficulty in reproducing Rähn's early results in the treatment of Cushing's Disease. This is probably because he used up to 4 re-treatments in some patients to achieve an overall success rate of 76%. Indeed the Stockholm group to which Rähn belongs did not achieve a high success rate when treating acromegaly.

There are lots of technical problems related to the primary treatment of these lesions with radiosurgery. This has been extensively discussed in the literature. Obviously as there is debate in this field it is inappropriate to state that one approach is optimal. However, it is suggested that today it is most reasonable to reserve radiosurgery for the treatment of cases where surgery has failed. This applies to roughly 20% of cases.

Chapter 13

Meningiomas

Three years ago there was still considerable debate about the appropriateness of radiosurgery for meningiomas. Moreover, amongst protagonists of this application of the method there was a diversity of dose schedules in use. This has all changed. There is a consensus that radiosurgery is an excellent method for treating residual or recurrent meningioma. Most people are cautious in using it as a primary treatment.

The doses most commonly in use today vary between 12 to 15 Gy to the margin. Knowledge has accumulated about the risk of complications. It would appear that non-basal meningiomas are associated with a much greater risk of complications than basal. The reason for this is not known, though the usual suggestion is that oedema may arise as a result of compression to superficial veins. This is however speculative at present.

There is one series where edge doses in excess of 18 Gy were associated with a much increased risk of radiation oedema. Moreover, in another series the risk of complications was related to tumour size: so that tumours with a mean diameter of more than 3 cm had a higher risk of complications than those which were smaller.

In most series to date the short term tumour control has been around 90 to 95%. The treatment seems to be of little value for malignant meningiomas.

Chapter 14

Cerebral Metastases

This has become a most popular indication. There is broad agreement that the method is most effective though there is debate as to how many metastases may be treated.

The main area of discussion for this indication relates to the use of prophylactic whole brain radiation. The literature supports arguments for and against its use. At present the matter is not resolved. In general the use of radiotherapy is popular in the USA and amongst oncologists. The use of radiosurgery as sole treatment is commoner amongst neurosurgeons and in Europe and the Far East.

Gliomas

Malignant Gliomas

The use of radiosurgery for malignant gliomas has been fairly disappointing with the possible exception of the Boston group who use a dedicated linear accelerator. Certainly, more studies need to be undertaken to clarify the role of the technique. There is no doubt that in some cases a dramatic palliation may be achieved. Thus, one of the main aims of research could be to determine which patients will react favourably before the treatment is administered.

Low Grade Gliomas

There are two papers (Ganz et al., Kihlström et al. – see reference list) which now indicate that the astrocytomas of childhood around the third ventricle have a most encouraging response to gamma knife radiosurgery.

Ocular Melanomas

There is a great deal of evidence that these could benefit from radiosurgery. However, at the present the treatment may be accompanied by a substantial number of complications, which while not dangerous are definitely uncomfortable. Many of the schedules used to date have not involved optimal conformational dose-planning. A multi-centre study is underway which aims to optimise and standardise the treatment parameters enabling all centres to achieve the results obtained so far only by the most expert practitioners.

Functional Disease

There has been an accelerating increase in the interest directed to this group of diseases. What follows is a short outline of the status quo.

Parkinson's Disease

There have been some encouraging early results. However, optimal dose-planning and consistency of results have not yet been achieved. It is the view of one of the world's leading experts, Professor Ohye, that the treatment has a role if used carefully. It is still under assessment and should not properly be used without adequate support in terms of neurological, neuro-physiological, neuro-psychological expertise. Moreover, it is widely held that the treatment of Parkinson's disease with the gamma knife should at present only be carried out by those who have previous experience of open thalamotomy.

Epilepsy

There has been an attempt to use radiosurgery in the treatment of epilepsy based on functional target localisation using a magneto-encephalogram. This is elegant work but has run into similar difficulties to those experienced by surgeons who perform cortical resections based on corticography. While there is optimism that technological advances will augment the usefulness of this method, it is true to say that to date it has been disappointing.

The treatment of anatomical targets has been far more promising. Thus, in Marseilles a sort of radiosurgical Yasargil's operation is performed with most promising early results and with a very careful cataloguing of the clinical radio-anatomical and physiological changes. This study has a near optimal design and its early results are most encouraging. (See Régis et al. in the reference list).

Another approach used by Whang et al. from Seoul in Korea is the application of gamma knife radiosurgery for patients with visible non-progressive lesions and intractable treatment resistant epilepsy. Also in these cases the early results have been most promising.

A multi-centre study is under way with the aim of quantifying the appropriate parameters for the use of the gamma knife in the treatment of epilepsy. However, even at this early stage it may be stated that its use based on anatomical localisation is promising.

Pain

There are a number of articles on this mostly from Professor Young from Seattle. He has achieved some interesting early results with thalamotomy for pain following thalamic lesions.

Trigeminal Neuralgia

There is now adequate material from the studies of Kondziolka and Young (see References) to support the contention that the Gamma Knife can be most effective in the treatment of this condition. Indeed, in the discussion section of his paper, Young goes so far as to indicate that a re-evaluation of treatment priorities may be necessary.

Both these authors emphasize that the results they record are as yet early. Even so, the maintenance of an early good response occurs in a satisfyingly high number of cases.

Psychosurgery

There was a meeting held in New York in June 1996 which decried the use of psychosurgery. The arguments were emotional and based more on theory than consideration of avoidable human suffering. The author would like to repeat what is indicated in Chapter 14 that the desuetude of psychosurgery under the pressure of political and religious forces is to be regretted.

Chapter 15

It is difficult to foresee the future. Thus it is pleasant that there is little to add to Chapter 15. The number of patients treated world-wide is over 50,000. There are nearly 90 units in operation.

The need for more knowledge remains urgent. However, it is also encouraging to note from the above appendix that substantial relevant progress has been made. Yet the task is vast and it is to be hoped that more and more centres will approach the work of radiosurgery in the spirit of Leksell. If they do then more and more centres will attack their labours with care and thoroughness and will honestly lay out the results of their efforts for all to read.

References

Radiosurgery Basics

1. Adams RD. The Neuropathology of Radiosurgery. Stereotact Funct Neurosurg 1991; 57: 82–86
2. Alexander III E, Siddon RL, Loeffler JS. The acute onset of nausea and vomiting following stereotactic radiosurgery: correlation with total dose to area postrema. Surg Neurol 1989; 32: 40–44
3. Altschuler EM, Lunsford LD, Flickinger JC. Radiobiological models for radiosurgery. Surg Forum 1990; 41: 506–508
4. Altschuler E, Lunsford LD, Kondziolka D, Wu A, Maitz AH, Sclabassi R, Martinez J, Flickinger JC. Radiobiologic models for radiosurgery. In: Lunsford LD (ed) Stereotactic Radiosurgery 1992; Philadelphia: WB Saunders Company: pp 61–77
5. Andersson B, Larsson B, Leksell L, Mair W, Rexed B, Sourander P, Wennerstrand J. Histopathology of late local radiolesions in the goat brain. Acta Radiol Ther Phys Biol 1970; 9: 385–394
6. Anniko M. Early morphological changes following gamma radiation. Acta Pathol Microbiol Immunol Scand 1981; 89: 113–124
7. Aristizabal SA, Boone JLM, Laguna JF. Endocrine factors influencing radiation injury to central nervous tissue. Int J Radiat Oncol Biol Phys 1979; 5: 349–353
8. Arndt J, Backlund EO, Larsson B, Leksell L, et al. Stereotactic irradiation of intracranial structures. Physical and biological considerations. Int J Dermatol 1979; 12: 81–92
9. Arnold A, Bailey P, Harvey RA, et al. Changes in the central nervous system following irradiation with 23 MeV x-rays from the betatron. Radiology 1954; 62: 37–46
10. Barnett GH. Evolution and Organization of a Regional Gamma Knife Center. Stereotact Funct Neurosurg (Suppl) 1996: 66; 365–369
11. Berg NO, Lindgren M. Dose factors and morphology of delayed radiation lesions of the middle ear in rabbits. Acta Radiol Stockh 1961; 56: 305–319
12. Berg NO, Lindgren M. Relation between field size and tolerance of rabbit's brain to roentgen irradiation (200 kV) via a slit-shaped field. Acta Radiol Stockh 1963; 1: 147
13. Bergström M, Greitz T, Steiner L. An approach to stereotaxis radiography. Acta Neurochir (Wien) 1980; 54: 157–165

14. Berk HW, Agarwal SK. Quality assurance of Leksell Gamma Units. Stereotact Funct Neurosurg 1991; 57: 106–112

15. Berk HW, Larner JM, Spaulding C, Agarwal SK, Scott MR, Steiner L. Extracranial absorbed doses with Gamma Knife radiosurgery. Stereotact Funct Neurosurg 1993; 61: 164–172

16. Bodis S, Alexander III E, Kooy H, Loeffler JS. The prevention of radiosurgery-induced nausea and vomiting by ondansetron: evidence of a direct effect on the central nervous system chemoreceptor trigger zone. Surg Neurol 1994; 42: 249–252

17. Boldrey E, Sheline G. Delayed transitory clinical manifestations after radiation treatment of intracranial tumors. Acta Radiol 1967; 5: 5–10

18. Bradshaw JD. Special report. The stereotactic radiosurgery unit in Sheffield. Clin Radiol 1986; 37: 277–279

19. Brahme A. Physical and radiobiologic aspects in the optimum choice of radiation modality. Acta Radiol Oncol 1982; 21: 469–479

20. Brahme A. Optimization of conformation and general moving beam radiation therapy techniques. In: Bruinvis et al. (ed) Proc 9th ICCR 1987; Amsterdam: Elsevier North-Holland: pp 227–230

21. Brahme A. Optimization of conformation and general moving beam radiation therapy techniques. Radiother Oncol 1988; 12: 129–140

22. Brismar J, Robertson GH, Davis KR. Radiation necrosis of the brain. Neuroradiological considerations with computed tomography. Neuroradiology 1976; 12: 109–113

23. Brown BW, Thompson JR, Barkley TB, et al. Theoretical considerations of dose rate factors influencing radiation strategy. Radiology 1973; 110: 97

24. Carbini CH, Goodman ML, Jones NE, Ford C. The use of magnetic resonance imaging in performing stereotactic surgery. In: Lunsford LD (ed) Stereotactic Radiosurgery Update 1992; New York: Elsevier: pp 67–72

25. Caveness WF. Pathology of radiation damage to the normal brain of the monkey. Natl Cancer Inst Monogr 1977; 46: 57–76

26. Chasan CB, Goetsch S, Ott K: Radiosurgery for Pineal Tumors: Is Biopsy Indicated? Stereotact Funct Neurosurg (Suppl) 1996: 66; 157–163

27. Chen GTY. Dose volume histograms in treatment planning. Int J Radial Oncol Biol Phys 1988; 14: 1319–1320

28. Coffey RJ, Lunsford LD. Stereotactic radiosurgery using the 201 Cobalt 60 Gamma Unit. In: Friedman WA (ed) Stereotactic Surgery Neurosurgery Clinics of North America 1990; Philadelphia: WB Saunders Company: pp 933–954

29. Coffey RJ, Lunsford LD, Flickinger JC. The role of radiosurgery in the treatment of malignant brain tumors. Neurosurgical Clinics of North America 1992; 3: 231–244

30. Coffey CW, Sanders M, Cashon K, Miller R, Walsh J, Patel P. A tissue equivalent phantom for stereotactic radiosurgery localization and dose verification. Stereotact Funct Neurosurg (Suppl) 1993; 61: 130–141

31. Dahlin H, Larsson B, Leksell L, Rosander K, Sarby B, Steiner L. Influence of absorbed dose and field size on the geometry of the radiation-surgical brain lesion. Acta Radiol Ther Phys Biol 1975; 14: 139–145

32. Dahlin H, Sarby B. Destruction of small intracranial tumors with co–60 gamma radiation. Physical and technical considerations. Acta Radiol Ther Phys Biol 1975; 14: 209–227

33. Dezymala RE, Mohan R, Brewster L, Chu J, Goitein M, Harms W, Urie M. Dose-volume histograms. Int J Radiat Oncol Biol Phys 1991; 21: 71–78

34. Di Lorenzo N, Nolletti A, Palma L. Late cerebral radionecrosis. Surg Neurol 1978; 10: 281–290

35. Dugger GS, Stratford JG, Buchard J. Necrosis of the brain following roentgen irradiation. Am J Roentgenol 1954; 72: 953–960

36. Edwards MS, Wilson CB. Treatment of radiation necrosis. In: Gilbert HA, Kagan AR (eds) Radiation Damage to the Nervous System. A Delayed Therapeutic Hazard 1980; New York: Raven Press: pp 129–143

37. Ellis F. Dose, time and fractionation. A clinical hypothesis. Clin Radiol 1969; 20: 1–7

38. Epstein ME, Lindquist C. Cost accounting the Gamma Knife. Stereotact Funct Neurosurg (Suppl) 1993; 61: 6–10

39. Ericson K, Söderman M, Maurincomme E, Lindquist C: Clinical Experience of Stereotaxic Digital Subtraction Angiography with Distortion Correction Software. Stereotact Funct Neurosurg (Suppl) 1996: 66; 63–70

40. Eyster EF, Nielsen SL, Sheline GE, et al. Cerebral radionecrosis simulating a brain tumor. J Neurosurg 1974; 39: 267–271

41. Flickinger JC. An integrated logistic formula for prediction of complications from radiosurgery. Int J Radiat Oncol Biol Phys 1989; 17: 879–885

42. Flickinger JC, Lunsford D, Kondziolka D. Dose prescription and dose-volume effects in radiosurgery. In: Lunsford LD (ed) Stereotactic Radiosurgery 1992; Philadelphia: WB Saunders Company: pp 51–59

43. Flickinger JC, Lunsford LD, Kondziolka D. Dose prescription and dose-volume effects in radiosurgery. Neurosurgical Clinics of North America 1991; 3: 51–59

44. Flickinger JC, Lunsford LD, Kondziolka D. Dose-volume considerations in radiosurgery. Stereotact Funct Neurosurg 1991; 57: 99–105

45. Flickinger JC, Lunsford LD, Kondziolka D, Maitz A. Potential human error in setting stereotactic coordinates for radiosurgery: implications for quality assurance. Int J Radiat Oncol Biol Phys 1993; 27: 397–401

46. Flickinger JC, Lunsford LD, Wu A, Kalend A. Predicted dose-volume isoeffect curves for radiosurgery with the 60 Co Gamma Unit. Acta Oncol 1991; 30: 363–367

47. Flickinger JC, Lunsford LD, Wu A, Kalend AM, Lindner G. Multiple

shot dosimetry for the gamma knife. Int J Radiat Oncol Biol Phys 1988; 15: S1:161

48. Flickinger JC, Lunsford LD, Wu A, Maitz A, Kalend AM. Treatment planning for Gamma Knife radiosurgery with multiple isocenters. Int J Radiat Oncol Biol Phys 1990; 18: 1495–1501

49. Flickinger JC, Maitz A, Kalend AM, Lunsford LD, Wu A. Treatment volume shaping with selective beam blocking using the Leksell Gamma Unit. Int J Radiat Oncol Biol Phys 1990; 19: 783–789

50. Flickinger JC, Schnell MC, Larson DA. Estimation of complications for linear accelerator radiosurgery with the integrated logistic formula. Int J Radiat Oncol Biol Phys 1990; 19: 143–148

51. Flickinger JC, Steiner L. Radiosurgery and the double logistic product formula. Radiother Oncol 1990; 17: 229–237

52. Foltz EL, Holyoke JB, Heyl HL. Brain necrosis following x-ray therapy. J Neurosurg 1953; 10: 423–429

53. Fu KK, Philips TL, Kane LJ, et al. Tumor and normal tissue response to irradiation in vivo. Variation with decreasing dose rates. Radiology 1975; 114: 709–716

54. Ganz JC. Gamma Knife Surgery. A Guide for Referring Physicians. 1993; Wien, New York: Springer-Verlag

55. Glauser TA, Packer RJ. Cognitive deficits in long term survivors of childhood brain tumors. Childs Nerv Syst 1991; 7: 2–12

56. Goetsch SJ, Hardy T, Hodgens D, Lizarraras A, Ott K, Scharfen C, Tung H. The Open Gamma Knife Center Concept. Stereotact Funct Neurosurg (Suppl) 1996: 66; 296–301

57. Graeb DA, Steinbok P, Robertson WD. Transient early computed tomographic changes mimicking tumor progression after brain tumor irradiation. Radiology 1982; 144: 813–817

58. Greitz T, Lax I, Bergström M, Arndt J, Berggren BM, Blomgren H. Stereotactic radiation therapy of intracranial lesions. Acta Radiol 1986; 25: 81–89

59. Guo WY, Chu WC, Wu MC, Chung WY, Gwan WP, Lee YL, Pan HC, Chang CY. Imaging Evaluation of Systematic Accuracy in MR-Guided Gamma Knife Surgery. Stereotact Funct Neurosurg (Suppl) 1996: 66; 85–92

60. Guy J, Mancuso A, Beck R, Moster M, Sedwich LA, Quisling RG, Thoton AL, Protzko EE, Schiffman J. Radiation-induced optic neuropathy: A magnetic resonance imaging study. J Neurosurg 1991; 74: 426–432

61. Hall EJ. Radiation dose-rate. A factor of importance in radiobiology and radiotherapy. Br J Radiol 1972; 45: 81–97

62. Hall EJ. Radiobiology for the Radiologists 1978; New York: Harper & Row

63. Hall EJ. Factors that Modify Radiobiological Response. In: Lunsford D. L. (ed) Stereotactic Radiosurgery Update; New York: Elsevier Scientific Publications; pp 11–18

64. Haymaker W, Ibrahim MZ, Miquel T, et al. Delayed radiation effects in the brains of monkeys exposed to x-rays. J Neuropathol Exp Neurol 1968; 27: 50–79

65. Helenowski TK, Pothiawala B. Role of the Gamma Knife in the Treatment of Large Lesions (Chicago, Ill) (Suppl 1). Stereotact Funct Neurosurg 1993; 61: 103–115

66. Hirato M, Hirato J, Zama A, Inoue H, Ohye C, Shibazaki T, Andou Y. Radiobiological Effects of Gamma Knife Radiosurgery on Brain Tumor Studied in Autopsy and Surgical Specimens. Stereotact Funct Neurosurg 1996 66; (Suppl) 4–16

67. Houdek PV, Fayos JV, Van Buren JM, Ginsberg MS. Stereotaxic radiotherapy technique for small intracranial lesions. Med Phys 1985; 12: 469–472

68. Hudgins WR. What is radiosurgery? Neurosurg 1988; 23: 272

69. Hudgins WR. Gamma Knife radiosurgery: Brain surgery without an incision. Tex Med 1993; 89: 64–68

70. Hudgins WR. Neurosurgery update – Gamma Knife radiosurgery. Dallas Med J 1994; 2: 69–73

71. Kalend AM, Wu A, Maitz AH, Saibon F, Yum GO, Izadbaksh MM. Dose distribution computation in stereotactic radiosurgery with a Gamma Knife (abstract). Med Phys 1992; 19: 791

72. Kihlström L, Guo W, Lindquist C, Mindus P. Radiobiology of radiosurgery for retractory anxiety disorders. Neurosurg 1995; 36: 294–302

73. Kingsley DP, Kendall BE. CT of the adverse effects of therapeutic radiation of the central nervous system. AJNR Am J Neuroradiol 1981; 2: 453–460

74. Kondziolka D, Lunsford LD, Claassen D, Maitz AH, Flickinger JC. Radiobiology of radiosurgery, Part 1. The normal rat brain model. Neurosurg 1992; 31: 271–279

75. Kondziolka D, Lunsford LD, Coffey RJ, Bissonette DJ, Flickinger JC. Stereotactic radiosurgery of angiographically occult vascular malformations: indications and preliminary experience. Neurosurg 1990; 27: 892–899

76. Kondziolka D, Lunsford LD, Flickinger JC. The role of radiosurgery in the management of chordomas and chondrosarcomas of the cranial base. Neurosurg 1991; 29: 38–46

77. Kondziolka D, Lunsford LD, Flickinger JC. Intraparenchymal brain stem radiosurgery. Neurosurg Clin N Am 1993; 4: 469–479

78. Kondziolka D, Somaza S, Flickinger J, Claassen D, Lunsford LD. Cerebral radioprotective effects of high-dose pentobarbital evaluated in an animal radiosurgery model. Neurol Res 1994; 16: 456–459

79. Lampert PW, Davis RL. Delayed effects of radiation on the human central nervous system. 'Early' and 'late' delayed reactions. Neurology 1964; 14: 912–917

80. Larner JM, Berk HW, Agarwal SK, Steiner L. The Dosimetric Consequences of Weighted Fields Using the Same Isocenter in Radiosurgery (Suppl 1). Stereotact Funct Neurosurg 1993; 61: 142–150

81. Larson DA. Introduction of radiosurgery. Neurosurg Clin N Am 1992; 1: 897–908

82. Larson DA, Bova F, Eisert D, Kline R, Loeffler J, Lutz W, Metha M, Palta J, Schewe K, Schultz C, Shaw E, Wilson F, Lunsford LD, Alexander E, Chapman P, Coffey R, Friedman W, Harsh G, Maciunas R, Olivíer A, Steinberg G, Walsh J. Consensus statement on stereotactic radiosurgery quality improvement. Int J Radiat Oncol Biol Phys 1993; 28: 527–530

83. Larson DA, Flickinger JC, Loeffler JS. The Radiobiology of Radiosurgery. Int J Radiat Oncol Biol Phys 1993; 25: 557–561

84. Larson DA, Flickinger JC, Loeffler JS. Stereotactic radiosurgery: Techniques and results. Principles & Practice of Oncology 1993; 7: 1–13

85. Larsson B. Blood vessel changes following local irradiation of the brain with high energy protons. Acta Soc Med Ups 1960; 65: 61–71

86. Larsson B, Leksell L, Rexed B, Sourander P. Effect of high energy protons on the spinal cord. Acta Radiol 1959; 51: 52–64

87. Larsson B, Leksell L, Rexed B, Sourander P, Mair W, Andersson B. The high energy proton beam as a neurosurgical tool. Nature 1958; 182: 1222–1223

88. Larsson B, Lidén K, Sarby B. Irradiation of small structures through the intact skull. Acta Radiol Ther Phys Biol 1974; 13: 512–534

89. Lax I, Brahme A, Andreo P. Electron beam dose planning using Gaussian beams; improved radial dose profiles. Acta Radiol Oncol Suppl 1983; 364: 49–59

90. Leksell DG. Stereotactic radiosurgery. Present status and future trends. Neurol Res 1987; 9: 60–68

91. Leksell DG. Special stereotactic techniques. Stereotactic radiosurgery. In: Heilbrun MP (ed) Concepts in Neurosurgery 1988; Baltimore: Williams & Wilkins: pp 195–209

92. Leksell DG. Stereotactic radiosurgery: Current status and future trends. Stereotact Funct Neurosurg 1993; 61: 1–5

93. Leksell DG, Arndt J. Risks of Radiosurgery (letter). Neurosurg 1989; 25: 480

94. Leksell L. A stereotaxic apparatus for intracerebral surgery. Acta Chir Scand 1949; 99: 229–233

95. Leksell L. The stereotaxic method and radiosurgery of the brain. Acta Chir Scand 1951; 102: 316–319

96. Leksell L. Cerebral radiosurgery (abstract). Proc 2nd Int Congr Neurol Surg: 1961; Washington DC: pp 7

97. Leksell L. Stereotactic radiosurgery. J Neurol Neurosurg Psychiat 1983; 46: 767–803

98. Leksell L, Herner T, Leksell DG, Persson B, Lindquist C. Visualization of stereotactic radiolesions by nuclear magnetic resonance. J Neurol Neurosurg Psychiat 1985; 48: 19–20

99. Leksell L, Herner T, Lidén K. Stereotaxic radiosurgery of the brain. Report of a case. Kungl Fysiograf Sällsk Lund Förhandl 1955; 25: 1–10

100. Leksell L, Larsson B, Andersson B, Rexed B, Sourander P, Mair W. Lesions in the depth of the brain produced by a beam of high energy protons. Acta Radiol 1960; 54: 251–264

101. Leksell L, Larsson B, Rexed B. The use of high energy protons for cerebral surgery in man. Acta Chir Scand 1963; 125: 1–7

102. Lindgren M. On tolerance of brain tissue and sensitivity of brain tumors to irradiation. Acta Radiol (Stockh) [Suppl] 1958; 170: 1–73

103. Lindquist C, Kondzioka D, Loeffler JS. Advances in Radiosurgery. Wien, New York: 1994; Springer-Verlag

104. Llena JF, Cespedes G, Hirano A, et al. Vascular alterations in delayed radiation necrosis of the brain. An electron microscopical study. Arch Pathol Lab Med 1976; 100: 531–534

105. Lowenberg-Scharenberg K, Bassett RC. Amyloid degeneration of the human brain following x-ray therapy. J Neuropathol Exp Neurol 1950; 9: 93–103

106. Lunsford LD. The Gamma Knife. Brain surgery without an incision. 'Frontiers of Medicine'. Hospital Physician 1988; 28–36

107. Lunsford LD. Current worldwide role of Gamma Knife stereotactic radiosurgery. In: Lunsford LD (ed) Stereotactic Radiosurgery Update 1992; New York: Elsevier; pp 31–35

108. Lunsford LD. Stereotactic radiosurgery: At the threshold or at the crossroads. Neurosurg 1993; 32: 811–816

109. Lunsford LD, Altschuler EM, Flickinger JC, Wu A, Martinez AJ. In vivo biological effects of stereotactic radiosurgery: A primate model. Neurosurg 1990; 27: 373–382

110. Lunsford LD, Flickinger JC, Coffey RJ. Stereotactic Gamma Knife radiosurgery as an alternative to intracranial microsurgery. Initial North American experience in 207 patients. Arch Neurol 1990; 47: 169–175

111. Lunsford LD, Flickinger JC, Lindner GJ, Maitz A. Stereotactic radiosurgery of the brain using the first United States 201 cobalt-60 source Gamma Knife. Neurosurg 1989; 24: 151–159

112. Lunsford LD, Kondziolka D, Flickinger JC. Stereotactic radiosurgery: Current spectrum and results. Clin Neurosurg 1992; 38: 405–444

113. Lunsford LD, Maitz A, Lindner G. The first U.S. 201 source cobalt 60 Gamma Unit for radiosurgery. Appl Neurophysiol 1988; 50: 253–256

114. Lunsford LD, Marion H. Gamma Knife stereotactic radiosurgery for brain tumors and vascular malformations. Experience in 207 consecutive patients. Hospimedica (Vol VII) 1989; 5: 40–49

115. Maciejewski B, Taylor JM, Withers HR. Alpha/beta value and the importance of size of dose per fraction for later complications in the supraglottic larynx. Radiother Oncol 1986; 7: 332–336

116. Mair W, Rexed B, Sourander P. Histology of the surgical radiolesion in the human brain as produced by high energy protons. Radiat Res Suppl 1967; 7: 394–389

117. Maitz AH, Lunsford LD, Wu A, Lindner G, Flickinger JC. Shielding requirements, on-site loading and acceptance testing of the Leksell Gamma Knife. Int J Radiat Oncol Biol Phys 1990; 18: 469–476

118. Major O, Kemeny AA, Forster DMC, Jakubowski J, Morice AH: In Vitro

Contractility Studies of the Rat Middle Cerebral Artery after Stereotactic Gamma Knife Radiosurgery. Stereotact Funct Neurosurg 1996; (Suppl) 66: 17–28

119. Mandybur TI, Gore I. Amyloid in the late post-irradiation necrosis of brain. Neurology 1969; 19: 983–992

120. Marks LB, Spencer DP. The influence of volume on the tolerance of the brain to radiosurgery. J Neurosurg 1991; 75: 177–180

121. Marra A, Guiffre R. Late cerebral radionecrosis. Eur Neurol 1968; 1: 234–246

122. Martins AN, Johnston JS, Henry JM, et al. Delayed radiation necrosis of the brain. J Neurosurg 1977; 47: 336–345

123. Mason DLD, Beddar AS, O'Brien PF. Dual–mode computerized linear accelerator for stereotactic radiosurgery (abstract). Med Phys 1992; 19: 789

124. Matsumura H, Ono H, Mori K. Delayed radiation necrosis. J Electron Microsc 1974; 6: 121–134

125. McDonald LW, Hayes TL. The role of capillaries in the pathogenesis of delayed radionecrosis of the brain. Am J Pathol 1967; 50: 745–764

126. McLaughlin WL, Soares CG. Dosimetry and Field Characteristics of Gamma Ray Stereotactic Radiosurgery Units Using Radiochromic Films (Abstract). Med Phys 1992; 19: 791

127. Meisberger LL, Keller RJ, Shale LR. The effective attenuation in water of the gamma rays, gold 198, iridum 192, cesium 137, radium 226, cobalt 60. Radiology 1968; 90: 953–958

128. Mikhael MA. Radiation necrosis of the brain. Correlation between computed tomography, pathology and dose distribution. J Comput Assist Tomogr 1978; 2: 71–80

129. Mikhael MA. Dosimetric consideration in the diagnosis of radiation necrosis of the brain. In: Gilbert HA, Kagan AR (eds) Radiation Damage to the Nervous System: A Delayed Therapeutic Hazard 1980; New York: Raven Press; pp 59–92

130. Nedzi LA, Kooy H, Alexander III E, Gelman RS, Loeffler JS. Variables associated the the development of complications of radiosurgery of intracranial tumors. Int J Radiat Oncol Biol Phys 1991; 21: 591–599

131. Nilsson A, Wennerstrand J, Leksell DG, Backlund EO. Stereotactic gamma irradiation of basiliar artery in cat. Preliminary experiences. Acta Radiol Oncol 1978; 17: 150–160

132. Ott K. A Comparison of Craniotomy Charges and Gamma Knife Charges in a Community Based Gamma Knife Center. Stereotact Funct Neurosurg (Suppl) 1996; 66: 357–364

133. Pay NT, Carella RJ, Lin JR, et al. The usefullness of computed tomography during and after radiation therapy in patients with brain tumors. Radiology 1976; 121: 79–83

134. Phelps ME, Hoffmann EJ, Ter-Gossian MM. Attenuation coefficients of various body tissues, fluids and lesions at photon energies of 18–136 KEV. Radiology 1975; 117: 573–583

135. Reinhold HS. The influence of radiation on blood vessels and circulation. Radiat Res 1974; 10: 1–198

136. Reinhold HS, Busiman GH. Repair of radiation damage to capillary endothelium. Br J Radiol 1975; 48: 727

137. Rider WD. Radiation damage to the brain. A new syndrome. J Can Assoc Radiol 1963; 14: 67–69

138. Rottenberg DA, Horten B, Kim JH, et al. Progressive white matter destruction following irradiation of an extracranial neoplasm. Ann Neurol 1980; 8: 76–78

139. Sanders M, Sayeg J, Coffey C, Patel P, Walsh J. Beam profile analysis using GafChromic films. Stereotact Funct Neurosurg (Suppl) 1993; 61: 124–129

140. Sarby B. Cerebral radiation surgery with narrow gamma beams. Physical experiments. Acta Radiol Ther Phys Biol 1974; 13: 425–444

141. Scholz W, Hsu YK. Late damage from roentgen irradiation of the human brain. Arch Neurol Psychiatry 1938; 40: 928–936

142. Shaw E, Kline R, Gillin M, Souhami L, Hirschfeld A, Dinapoli R, Martin L. Radiation therapy oncology group: Radiosurgery quality assurance guidelines. Int J Radiat Oncol Biol Phys 1993; 27: 1231–1239

143. Sheline GE, Wara WM, Smith V. Therapeutic irradiation and brain injury. Int J Radiat Oncol Biol Phys 1980; 6: 1215–1228

144. Smith V, Verhey L, Jones E, Lyman J. Consequences to the patient in the event of hydraulic unit failure. Stereotact Funct Neurosurg (Suppl) 1993; 61: 173–177

145. Smith V, Verhey L, Wara W, Harsh G, Larson D. Beta test site report: Gamma Plan. Stereotact Funct Neurosurg (Suppl) 1993; 61: 116–123

146. Terahara A. Gamma Knife. Gan To Kagaku Ryoho 1993; 20: 2133–2142

147. Thorsen FA, Ganz JC. Dose planning with the Leksell Gamma Knife: The Effect on Dose Volume of More Than One Shot at the Same Target Point (Bergen) (Suppl 1). Stereotact Funct Neurosurg 1993; 61: 151–163

148. Tsuzuki T, Tsunoda S, Sakaki T, Konishi N, Hiasa Y, Nakamura M, Yoshino E: Tumor Cell Proliferation and Apoptosis associated with Gamma Knife Effect. Stereotact Funct Neurosurg 1996 (Suppl.) 66; 39–48

149. Valk PE, Dillon WP. Radiation injury of the brain. AJNR Am J Neuroradiol 1991; 12: 45–62

150. Van der Kogel AJ. Central nervous system radiation system injury in small animal models. In: Gutin PH, Lerbel SA, Sheline GE (eds) Radiation Injury to the Nervous System 1991; New York: Raven Press; pp 91–111

151. Wachowski TJ, Chenault H. Degenerative effects of large doses of roentgen rays on the human brain. Radiology 1945; 45: 227–246

152. Walton L, Bomford CK, Ramsden D. The Sheffield stereotactic radiosurgery unit. Physical characteristics and principles of operation. Br J Radiol 1987; 60: 897–906

153. Walton L, Griffiths LM, Hampshire A. Recomissioning of the Sheffield Gamma Knife Unit Following a Source Change (Sheffield) (Suppl 1). Stereotact Funct Neurosurg 1993; 61: 179–185

154. Walton L, Hampshire A, Forster DMC, Kemeny AA. Accuracy of Stereotactic Localisation Using MRI: A Comparison between 2D and 3D studies. Stereotact Funct Neurosurg (Suppl) 1996: 66; 49–56

155. Whang J. Gamma Knife and its clinical application. Chung Hua Wai Ko Tsa Chih 1992; 30: 690–692

156. Whitmore GF, Gulyas S, Botond J. Radiation sensitivity throughout the cell cycle and its relationship to recovery. In: Cellular Radiation Biology 1965; Baltimore:Williams & Wilkins: pp 423–431

157. Wilson GH, Byfield J, Hanafee WN. Atrophy following radiation therapy for CNS neoplasms. Acta Radiol (Ther Ther Phys Biol) [Stockh] 1972; 11: 361–368

158. Wu A. Physics and dosimetry of the Gamma Knife. In: Lunsford LD (ed) Stereotactic Radiosurgery 1992; Philadelphia: WB Saunders Company; pp 35–50

159. Wu A. The future of Gamma Knife approach in stereotactic radiosurgery (abstract). Med Phys 1992; 19: 855

160. Wu A, Kalend AM. A new algorithm for Gamma Knife treatment planning system. In: Lunsford LD (ed) Stereotactic Radiosurgery Update 1992; New York: Elsevier; pp 217–220

161. Wu A, Lindner G, Maitz A, Kalend AM, Lunsford LD, Flickinger JC, Bloomer WD. Physics of Gamma Knife approach on convergent beams in stereotactic radiosurgery. Int J Radiat Oncol Biol Phys 1990; 18: 941–949

162. Wu X, Ting JY, Markoe AM, Landy HJ, Fiedler JA, Russell J. Stereotactic Dose Computation and Plan Optimization Using Convolution Theorem Dose Computation. Stereotact Funct Neurosurg (Suppl) 1996: 66; 302–308

163. Yamamoto M, Jimbo M, Ide M, Umebara Y, Hagiwara S, Takakura K, Hirai T. Is Unchanged Tumor Volume after Radiosurgery a Measure of the Outcome? Stereotact Funct Neurosurg (Suppl) 1996: 66; 231–239

164. Yamamoto M, Lindquist C. Determination of target point magnification rate on stereotactic angiography for radiosurgery. Technical note. Neurol Med Chir (Tokyo) 1993; 33: 391–394

165. Yoshii Y, Phillips TL. Late vascular effects of whole brain X-irradiation in the mouse. Acta Neurochir (Wien) 1982; 64: 87–102

Vascular Malformations

1. 1. Altschuler EM, Lunsford LD, Coffey RJ, Bissonette DJ, Flickinger JC. Gamma Knife Radiosurgery for Intracranial Arteriovenous Malformations in Childhood and Adolescence. Pediatr Neurosci 1989; 15: 53–61

2. Backlund EO, Arndt J, Dahlin H, Greitz T, Leksell L, Steiner L. Radiosurgery in arteriovenous malformations I. Technique. In: Carrea R, Le Vey D (eds) Neurological Surgery 1977; Amsterdam–Oxford: Excerpta Medica; pp 162–167

3. Bunge H, Chinela A, Lemme-Plaghos L, Guevara J, Steiner L. Radiosurgical treatment of intracranial arteriovenous malformations with gamma knife. Neuroradiology 1991; 33: 206–208

4. Bunge HJ, Guevara JA, Chinela AB. Malformaciones Arteriovenosas: Experiencia con la Gamma Unit III en el Centro de Radiocirugia Neurologica del Sol. Rev Ven de Neurol y Neurocir 1990; 4: 15–19

5. Ciarmatori C, Tassinari CA, Calbucci F, Steiner L. Follow-up clinico e neuroradiologico in un caso di cavernoma cerebrale trattato mediante radiochirurgia stereotassica. Riv Neurobiologica 1989; 35: 137–140

6. Coffey RJ, et al. Stereotactic radiosurgical treatment of cerebral arteriovenous malformations. Gamma Unit Radiosurgery Study Group. Mayo Clin Proc 1995; 70: 214–22

7. Duma CM et al. Radiosurgery for vascular malformations of the brain stem. Acta Neurochir Suppl (Wien) 1993; 58: 92–7

8. Flickinger JC, Lunsford LD, Kondziolka D, Maitz AH, Epstein AH, Simons SR, Wu A. Radiosurgery and brain tolerance: An analysis of neurodiagnostic imaging changes after Gamma Knife radiosurgery for arteriovenous malformations. Int J Radiat Oncol Biol Phys 1992; 23: 19–26

9. Forster DMC, Kunkler OH, Hartland RI. Risk of cerebral bleeding from arteriovenous malformations in pregnancy: The Sheffield experience. Stereotact Funct Neurosurg (Suppl) 1993; 61: 20–22

10. Forster DMC, Steiner L, Håkansson S. Arteriovenous malformation of the brain. A long-term clinical study. J Neurosurg 1972; 37: 562–570

11. Guo WY. Radiobiological aspects of Gamma Knife radiosurgery for arteriovenous malformations and other non-tumoural disorders of the brain. Acta Radiol Suppl 1993; 338: 1–34

12. Guo WY, Lindquist C, Karlsson B, Kihlström L, Steiner L. Gamma Knife surgery of cerebral arteriovenous malformations: Serial MR imaging studies after radiosurgery. Int J Radiat Oncol Biol Phys 1993; 25: 315–323

13. Guo WY, Lindqvist M, Lindquist C, Ericson K, Nordell B, Karlsson B, Kihlström L. Stereotaxic angiography in Gamma Knife radiosurgery of intracranial arteriovenous malformations. AJNR Am J Neuroradiol 1992; 13: 1107–1114

14. Guo WY, Nordell B, Karlsson B, Söderman M, Lindqvist M, Ericson K, Franck A, Lax I, Lindquist C. Target delineation in radiosurgery for cerebral arteriovenous malformations. Assessment of the value of stereotaxic MR imaging and MR angiography. Acta Radiol 1993; 34: 457–463

15. Guo WY, Pan DHC, Liu RS, Chung WH, Shiau CY, Cheng SS, Chang CY, Chen KY, Yeh SH, Lee LS. Early irradiation effects observed on

magnetic resonance imaging and angiography, and positron emission tomography for arteriovenous malformations treated by Gamma Knife radiosurgery. Stereotact Funct Neurosurg 1995; 64: 258–269

16. Guo WY, Pan HC, Chung WY, Wang LW, Teng MMH: Do We Need Conventional Angiography? The Role of MR in Verifying Obliterated AVMs after Gamma Knife Surgery. Stereotact Funct Neurosurg (Suppl) 1996: 66; 71–84

17. Guo WY, Wikholm G, Karlsson B, Svendsen P, Ericson K. Combined embolization and Gamma Knife radiosurgery for cerebral arteriovenous malformations. Acta Radiol 1993; 34: 600–606

18. Karlsson B, Lax I, Söderman M, Kihlström L, Lindquist C. Prediction of Results Following Gamma Knife Surgery for Brain-Stem and other Centrally Located Arteriovenous Malformations in Relation to the Natural Course. Stereotact Funct Neurosurg (Suppl) 1996: 66; 260–269

19. Kawamoto S, Sasaki T, Takakura K, Aoki S, Terahara A, Akanuma A. The use of computed tomography (CT) as a reference for dose planning and treatment of arteriovenous malformations. In: Lunsford LD (ed) Stereotactic Radiosurgery Update 1992; New York: Elsevier; pp 145–151

20. Kemeny AA, Dias PS, Forster DMC. Results of stereotactic radiosurgery of arteriovenous malformations: an analysis of 52 cases. J Neurol Neurosurg Psychiat 1989; 52: 554–558

21. Kobayashi T, et al. Gamma knife treatment of AVM of the basal ganglia and thalamus. No To Shinkei 1996; 484: 351–6

22. Kondziolka D, Lunsford LD, Kanal E, Talagala L. Stereotactic magnetic resonace angiography for targeting in arteriovenous malformation radiosurgery. Neurosurg 1994; 35: 585–591

23. Leber KA, Aigner R, Nicoletti R, Fueger GF, Pendl G. Dynamic and Static Scintigraphic Evaluation of Cerebral Arteriovenous Malformations with Respect to Radiosurgical Treatment. Stereotact Funct Neurosurg (Suppl) 1996; 66: 269–277

24. Lindquist C, Guo WY, Karlsson B, Steiner L. Radiosurgery for venous angiomas. J Neurosurg 1993; 78: 531–536

25. Lindquist C, Steiner L, Blomgren H. Stereotactic radiation therapy of intracranial arteriovenous malformations. Acta Radiol (Stockh) [Suppl] 1986; 369: 610–613

26. Lindquist C, Steiner L. Stereotactic radiosurgical treatment of arteriovenous malformations. In: Lunsford LD (ed) Modern Stereotactic Radiosurgery 1988; Boston: Martinus Nijhoff; pp 491–505

27. Lunsford LD. Stereotactic heavy-charged-particle Bragg peak radiosurgery for the treatment of intracranial arteriovenous malformations in childhood and adolescence (comment). Neurosurg 1989; 24: 841

28. Lunsford LD. Stereotactic radiosurgery of intracranial arteriovenous malformations. In: Wilkens RH, Rengachary SS (eds) Neurosurgery Update 1990; New York: McGraw Hill; pp 175–185

29. Lunsford LD, Kondziolka D, Flickinger JC, Bissonette DJ, Jungreis CA, Maitz AH, Horton JA, Coffey RJ. Stereotactic radiosurgery for arteriovenous malformations of the brain. J Neurosurg 1991; 75: 512–524

30. Ogilvy CS. Radiation therapy for arteriovenous malformations: A review. Neurosurg 1990; 26: 725–735

31. Piovan E, Dal Sasso M, Urbani GP, Sartori R, Foroni R, Benati A. The Use of Digital Subtraction Angiography for Arteriovenous Malformations in Stereotactic Radiosurgery. Stereotact Funct Neurosurg (Suppl) 1996: 66; 57–62

32. Pollock BE, et al. Hemorrhage risk after stereotactic radiosurgery of cerebral arteriovenous malformations. Neurosurgery 1996; 38: 652–659

33. Pollock BE, et al. Patient outcomes after stereotactic radiosurgery for operable arteriovenous malformations. Neurosurgery 1994; 35: 1–7

34. Pollock BE, et al. Stereotactic radiosurgery for postgeniculate visual pathway arteriovenous malformations. J Neurosurg 1996; 84: 437–41

35. Pollock BE, Kondziolka D, Lunsford LD, Bissonette D, Flickinger JC. Repeat stereotactic radiosurgery of arteriovenous malformations: factors associated with incomplete obliteration. Neurosurg 1996; 38: 318–324

36. Pollock BE, Lunsford LD, Kondziolka D, Bissonette DJ, Flickinger JC. Stereotactic radiosurgery for postgeniculate visual pathway arteriovenous malformations. J Neurosurg 1996; 84: 437–441

37. Pollock BE, Lunsford LD, Kondziolka D, Maitz A, Flickinger JC. Patient outcomes after stereotactic radiosurgery for 'operable' arteriovenous malformations. Neurosurg 1994; 35: 1–8

38. Sadler LR, Jungreis CA, Lunsford LD, Trapanollo MM. Angiographic technique to precede Gamma Knife radiosurgery for intracranial arteriovenous malformations. AJNR Am J Neuroradiol 1990; 11: 1157–1161

39. Spaulding CA, Berk HW. Stereotaxic radiosurgery in the treatment of arteriovenous malformations. Appl Radiol 1989; 18: 11–14

40. Steiner L. Stereotactic radiosurgery with the cobalt 60 Gamma Unit in the surgical treatment of intracranial tumors and arteriovenous malformations. In: Schmidek HH, Sweet WH (eds) Operative Neurosurgical Techniques 1988; Orlando: Grune & Stratton; pp 515–529

41. Steiner L. Treatment of arteriovenous malformations by radiosurgery. In: Wilson C, Stein B (eds) Intracranial Arteriovenous Malformations 1984; Baltimore: Williams & Wilkins; pp 295–313

42. Steiner L, Greitz T, Backlund EO, Leksell L, Norén G, Rähn T. Radiosurgery in arteriovenous malformations of the brain. Undue effects. INSERM 1979; 12: 257–269

43. Steiner L, Greitz T, Forster DMC, Leksell L, Backlund EO. Stereotaxic radiosurgery for cerebral arteriovenous malformations. Report of a case. Acta Chir Scand 1972; 138: 459–464

44. Steiner L, Greitz T, Leksell L, Norén G, Rähn T, Backlund EO. Radiosurgery in intracranial arteriovenous malformations II. A follow-

up study. In: Carrea R, Le Vay D (eds) Neurological Surgery 1977; Amsterdam–Oxford: Excerpta Medica; pp 168–180

45. Steiner L, Leksell L, Forster DMC, Greitz T, Backlund EO. Stereotaxic radiosurgery in intracranial arteriovenous malformations. Acta Neurochir (Wien) [Suppl] 1974; 21: 195–209

46. Steiner L, Leksell L, Greitz T, Forster DMC, Backlund EO. Stereotaxic radiosurgery for cerebral arteriovenous malformations. Report of a case. Acta Chir Scand 1972; 138: 459–464

47. Steiner L, Leksell L, Greitz T, Forster DMC, Backlund EO. Stereotaxic radiosurgery for cerebral arteriovenous malformations. Yearbook of Neurology and Neurosurgery 1974; pp 452–454

48. Steiner L. Radiosurgery in cerebral arteriovenous malformations. In: Flamm E, Fein J (eds) Textbook of Cerebrovascular Surgery 1986; New York: Springer-Verlag; pp 1161–1215

49. Steiner L, Lindquist C. Radiosurgery in cerebral arteriovenous malformations. In: Tasker R (ed) Neurosurgery State of the Art Reviews, No. 2 1987; pp 329–336

50. Steiner L, Lindquist C, Adler JR, Torner JC, Alves W, Steiner M. Clinical outcome of radiosurgery for cerebral arteriovenous malformations. J Neurosurg 1992; 77: 1–8

51. Steiner L, Lindquist C, Cail W, Steiner M. Microsurgery and radiosurgery in brain arteriovenous malformations. J Neurosurg 1993; 79: 647–652

52. Tanaka T, et al. The comparison between adult and pediatric AVMs treated by gamma knife radiosurgery. No Shinkei Geka 1995; 23: 773–7

53. Tanaka T, Kobayashi T, Kida Y, Oyama H, Niwa M. The Comparison between Adult and Pediatric Arteriovenous Malformations treated by Gamma Knife Radiosurgery. Stereotact Funct Neurosurg (Suppl) 1996: 66; 288–296

54. Yamamoto M, et al. Gamma knife radiosurgery for cerebral arteriovenous malformations: an autopsy report focusing on irradiation-induced changes observed in nidus-unrelated arteries. Surg Neurol 1995; 44: 421–7

55. Yamamoto M, Jimbo M, Ide M, Lindquist C, Steiner L. Postradiation volume changes in Gamma Knife-treated cerebral arteriovenous malformations. Surg Neurol 1993; 40: 485–490

56. Yamamoto M, Jimbo M, Ide M, Tanaka N, Lindquist C, Steiner L. Long term follow-up of radiosurgically treated arteriovenous malformations in children: Report of nine cases. Surg Neurol 1992; 38: 95–100

57. Yamamoto M, Jimbo M, Kobayashi M, Yoyoda C, Ide M, Tanaka N, Lindquist C, Steiner L. Long-term results of radiosurgery for arteriovenous malformation: Neurodiagnostic imaging and histological studies of angiographically confirmed nidus obliteration. Surg Neurol 1992; 37: 219–230

58. Yamamoto M, Jimbo M, Lindquist C. Radiation-induced edema after radiosurgery for pontine arteriovenous malformation. A case report

and detection by magnetic resonance imaging. Surg Neurol 1992; 37: 15–21

59. Yamamoto Y, et al. Interim report on the radiosurgical treatment of cerebral arteriovenous malformations. The influence of size, dose, time, and technical factors on obliteration rate. J Neurosurg 1995; 83: 832–837

Acoustic Schwannoma

1a. Anniko M, Arndt J, Norén G. The human acoustic neurinoma in organ culture II. Tissue changes after gamma irradiation. Acta Otolaryngol (Stockh) 1981; 91: 223–235

1b. Anniko M, Norén G. The human acoustic neurinoma in organ culture I. Methodological aspects. Acta Otolaryngol (Stockh) 1981; 91: 47

2. Brackmann D, Kwartler JA. Treatment of acoustic tumors with radiotherapy (Commentary). Arch Otolaryngol Head Neck Surg 1990; 116: 161–162

3. Flickinger JC, Lunsford LD, Coffey RJ, Linskey ME, Bissonette DJ, Maitz AH, Kondziolka D. Radiosurgery of acoustic neurinomas. Cancer 1991; 67: 345–353

4. Flickinger JC, Lunsford LD, Kondziolka D, Coffey RJ. Radiosurgery of acoustic neuromas. Proc ASTROMiami Beach 1990

5. Flickinger JC, Lunsford LD, Linskey ME, Duma CM, Kondziolka D. Gamma Knife radiosurgery for acoustic tumors: Multivariate analysis of four year results. Radiat Oncol 1993; 27: 91–98

6. Forster DMC, Kemeny AA, Pathak A, Walton L. Radiosurgery: a minimally interventional alternative to microsurgery in the management of acoustic neuroma. Br J Neurosurg 1996; 10: 169–174

7. Ganz JC, Mathisen JR, Thorsen F, Backlund E-O. Acoustic neurinoma: Early results related to radiobiological variables. In: Lunsford LD (ed) Stereotactic Radiosurgery Update 1992; New York: Elsevier: pp 359–363

8. Helenowski TK. Gamma Knife radiosurgery of acoustic neuromas. Cinn Report 1994; 38–42

9. Hirato M, Inoue H, Nakamura M, Ohye C, Hirato J, Shibazaki T, Andou Y. Gamma knife radiosurgery in acoustic neurinoma: early effects and the preservation of hearing. Neuro Medico-Chir (Tokyo) 1995; 35: 737–741

10. Hirato M, Inoue HK, Nakamura M, Ohye C, Shibazaki T, Andou Y. Gamma knife radiosurgery in acoustic neurinoma: Early effects and the preservation of hearing. Acta Neurochir (Wien) 1993; 122: 165

11. Hirato M, Inoue H, Zama A, Ohye C, Shibazaki T, Andou Y. Gamma Knife Radiosurgery for Acoustic Schwannoma Effects of Low Radiation Dose and Functional Prognosis. Stereotact Funct Neurosurg (Suppl) 1996: 66; 134–141

12. Hirsch A, Norén G, Anderson H. Audiologic findings after stereotactic

radiosurgery in nine cases of acoustic neurinomas. Acta Otolaryngol (Stockh) 1979; 88: 155–160

13. Hirsch A, Norén G, Anderson H. Audiological findings after stereotactic radiosurgery in acoustic neurinomas. Acta Otolaryngol (Stockh) 1988; 106: 244–251

14. Hirsch A, Norén G. Audiological findings after stereotactic radiosurgery in acoustic neurinomas. Acta Otolaryngol (Stockh) 1988; 106: 244–251

15. Kamerer DB, Lunsford LD, Møller M. Gamma knife: An alternative treatment for acoustic neurinomas. Ann Otol Rhinol Laryngol 1988; 97: 631–635

16. Kondziolka D, Lunsford LD. Preservation of hearing in acoustic neurinoma surgery. J Neurosurg 1993; 78: 154–156

17. Kondziolka D, Lunsford LD, Coffey RJ, Flickinger JC. Cranial nerve preservation after stereotactic radiosurgery of acoustic neuroma. Surg Forum 1990; 41: 508–510

18. Kondziolka D, Lunsford LD, Coffey RJ. Cranial nerve preservation after stereotactic radiosurgery of acoustic neurinomas. Proc ACS San Francisco 1990

19. Laasonen EM, Troupp H. Volume growth rate of acoustic neurinomas. Neuroradiology 1986; 28: 203–207

20. Leksell L. A note on the treatment of acoustic tomors. Acta Chir Scand 1971; 137: 763–765

21. Linskey ME, Lunsford LD, Flickinger JC. Neuroimaging of acoustic nerve sheath tumors after stereotactic radiosurgery. AJNR 1991; 12: 1165–1175

22. Linskey ME, Lunsford LD, Flickinger JC. Radiosurgery for acoustic neurinomas: Early experience. Neurosurgery 1990; 26: 736–745

23. Linskey ME, Lunsford LD, Flickinger JC. Stereotactic radiosurgery for acoustic nerve sheath tumors. In: Lunsford LD (ed) Stereotactic Radiosurgery Update 1992; New York: Elsevier; pp 321–334

24. Linskey ME, Lunsford LD, Flickinger JC. Tumor control after stereotactic radiosurgery in neurofibromatosis patients with bilateral acoustic tumors. Neurosurgery 1992; 31: 829–839

25. Linskey ME, Lunsford LD, Flickinger JC, Kondziolka D. Stereotactic radiosurgery for acoustic tumors. Neurosurg Clin North Am 1992; 3: 191–205

26. Linskey ME, Lunsford LD, Flickinger JC, Kondziolka D. Stereotactic radiosurgery for acoustic tumors. In: Lunsford LD (ed) Stereotactic Radiosurgery 1992; Philadelphia: WB Saunders; pp 191–205

27. Linskey ME, Lunsford LD, Flickinger JC. Stereotactic radiosurgery for acoustic nerve sheath tumors. In: Lunsford LD (ed) Stereotactic Radiosurgery Update 1992; New York: Elsevier; pp 321–334

28. Linskey ME, Lunsford LD, Flickinger JC. Cranial nerve length predicts the risk of delayed facial and trigeminal neuropathies of the acoustic tumor treated with stereotactic radiosurgery. Int J Radiat Oncol Biol Phys 1993; 25: 227–233

29. Linskey ME, Martinez AJ, Kondziolka D, Flickinger JC, Maitz AH, Whiteside T, Lunsford LD. The radiobiology of human acoustic schwannoma xenografts after stereotactic radiosurgery evaluated in the subrenal capsule of athymic mice. J Neurosurg 1993; 78: 645–653

30. Lunsford LD, Coffey RJ, Bissonette DJ, Flickinger JC. Radiosurgery for acoustic neurinoma. Proc Harvard Radiosurgery Update Boston 1990

31. Lunsford LD, Goodman M. Stereotactic radiosurgery for acoustic neurinomas. Surg Forum 1988; 39: 505–507

32. Lunsford LD, Kamerer DB, Flickinger JC. Stereotactic radiosurgery for acoustic neurinomas. Arch Otolaryngol (Head Neck Surgery) 1990; 116: 907–909

33. Lunsford LD, Kamerer DB, Möller M. Gamma Knife: An alternative treatment for acoustic neurinomas. Ann Otol Rhinol Laryngol 1988; 97: 631–635

34. Lunsford LD, Kondziolka D, Flickinger JC. Radiosurgery as an alternative to microsurgery for the treatment of acoustic neurinomas. Clin Neurosurg 1990; 38: 619–634

35. Lunsford LD, Linskey ME. Stereotactic radiosurgery in the treatment of patients with acoustic tumors. Otolaryngol Clin North Am 1992; 25: 471–491

36. Nagano H, Tanohata K, Kato E, Nakayama S, Fujino H, Matsubara S. Dose Distribution and Shrinkage of Acoustic Neurinomas at 2 Years after Gamma Knife Treatment. Stereotact Funct Neurosurg (Suppl) 1996: 66; 146–156

37. Norén G. Gamma Knife radiosurgery in acoustic neurinomas. In: Haid CT (ed) Vestibular Diagnosis and Neuro-Otosurgical Management of the Skull Base 1991; Gräfelfing: Demeter Verlag; pp 43–47

38. Norén G, Arndt J, Hindmarch T, Hirsch A. Stereotactic radiosurgical treatment of acoustic neurinomas. In: Lunsford LD (ed) Modern Stereotactic Neurosurgery 1988; Boston: Martinus Nijhoff; pp 481–489

39. Norén G, Arndt J, Hindmarch T. Stereotactic radiosurgery in cases of acoustic neurinomas. Further experiences. Neurosurgery 1983; 13: 12–22

40. Norén G, Backlund EO, Grepe A, Leksell L. Stereotactic radiosurgical treatment of acoustic neurinomas. Acta Neurochir (Wien) 1979; 45: 337–338

41. Norén G, Backlund EO. Treatment of acoustic tumors by stereotactic radiosurgery. Proc 6th Int Congr Neurol Surg Sao Paolo 1977

42. Norén G, Greitz D, Hirsch A, Lax I. Gamma knife radiosurgery in acoustic neurinoma. In: Steiner L, Lindquist C, Forster D, Backlund E-O (eds) Radiosurgery: Baseline and Trends 1992; New York: Raven Press; pp141–148

43. Norén G, Greitz D, Hirsch A, Lax I. Gamma Knife radiosurgery in acoustic neurinomas. In: Tos M, Thomsen J (eds) Acoustic Neuroma 1992; Amsterdam: Kugler; pp 289–292

44. Norén G, Greitz A, Lax I. Gamma Knife surgery in acoustic tumors. Acta Neurochir (Wien) 1993; 58: 104–107

45. Norén G, Hirsch A, Mosskin M. Long-term efficacy of Gamma Knife radiosurgery in vestibular schwannomas (abstract). Acta Neurochir (Wien) 1993; 122: 164

46. Norén G, Leksell L. Stereotactic treatment of acoustic tumors. INSERM 1979; 12: 241–244

47. Ogunrinde OK, Lunsford LD, Flickinger JC, et al. Cranial nerve preservation after radiosurgery for small acoustic tumors. Arch Neurol 1995; 52: 73–79

48. Ogunrinde OK, Lunsford LD, Flickinger JC, Kondziolka D. Stereotactic radiosurgery for acoustic nerve tumors in patients with useful preoperative hearing: results at 2-year follow-up examination. J Neurosurg 1994; 80: 1011–1017

49. Ogunrinde OK, Lunsford LD, Flickinger JC, Kondziolka DS. Cranial nerve preservation after stereotactic radiosurgery for small acoustic tumors. Arch Neurol 1995; 52: 73–79

50. Pendl G, Ganz JC, Schröttner O, Eustacchio S. Acoustic Neurinomas with Macrocysts Treated with Gamma Knife Radiosurgery. Stereotact Funct Neurosurg (Suppl) 1996: 66; 103–111

51. Pollock BE, Kondziolka D, Flickinger JC, Maitz AM, Lunsford LD. Preservation of Cranial Nerve Function after Radiosurgery for Nonacoustic Schwannomas. Neurosurgery 1993; 33: 597–601

52. Pollock BE, Lunsford LD, Kondzioloka D, Flickinger JC, Bissonette DJ, Kelsey SF, Jannetta PJ. Outcome analysis of acoustic neuroma management: A comparison of microsurgery and stereotactic radiosurgery. Neurosurgery 1995; 36: 316–229

53. Seo Y, Fukuoka S, Nakagawara J, Takanashi M, Takahashi S, Suematsu K, Nakamura J. The Effect of Gamma Knife Radiosurgery on Acoustic Neurinomas: Assessment by 99mTc-DTPA-Human Serum Albumin SPECT and 201TlCl SPECT. Stereotact Funct Neurosurg (Suppl) 1996: 66; 93–102

54. Thomsen J, Tos M. Acoustic neuroma: Clinical aspects, audiovestibular assessment, diagnostic delay, and growth rate. Am J Otol 1990; 11: 12–19

55. Thomsen J, Tos M, Borgesen SE. Gamma Knife: Hydrocephalus as a complication of stereotactic radiosurgical treatment of an acoustic neuroma. Am J Otol 1990; 11: 330–333

56. Wallner KE, Sheline GE, Pitts LH, Wara WM, Davis RL, Boldray EB. Efficacy of irradiation for incompletely excised acoustic neurilemmomas. J Neurosurg 1987; 67: 858–863

57. Yamamoto M, Hagiwara S, Ide M, Jimbo M, Hirai T, Nakamura Y. Radiosurgery for acoustic neurinoma with rapid growth and relatively high staining indexes for PCNA and MIB-1: A case report. Neurol Med Chir (Tokyo) 1996; 36(4): 241–245

58. Yamamoto M, Norén G. Stereotactic radiosurgery in acoustic

neurinomas. No-Shinkei-Geka (Neuro Surg Tokyo) 1990; 18: 1101–1106

Meningiomas

1. Duma CM, Lunsford LD, Kondziolka D, Harsh GR, Flickinger JC. Stereotactic radiosurgery of cavernous sinus meningiomas as an addition or alternative to microsurgery. Neurosurgery 1993; 32: 1–8
2. Duma CM, Lunsford LD, Kondziolka D, Harsh IV GR, Flickinger JC. Stereotactic radiosurgery of cavernous sinus meningiomas as addition or alternative to microsurgery. Neurosurgery 1993; 32: 699–705
3. Firsching RP, Fischer A, Peters R, Thun F, Klug N. Growth rate of incidental meningiomas. J Neurosurg 1990; 73: 545–547
4. Ganz JC, Backlund EO, Thorsen FA. The results of Gamma Knife surgery of meningiomas, related to size of tumor and dose. Stereotact Funct Neurosurg 1993; 61 (Suppl 1): 23–9
5. Ganz JC, Schröttner O, Pendl G. Radiation-induced Edema after Gamma Knife Treatment for Meningiomas. Stereotact Funct Neurosurg (Suppl) 1996; 66: 129–133
6. Hudgins WR, Barker JL, Schwarz DE, Nichols TD. Gamma Knife Treatment of 100 Consecutive Meningiomas. Stereotact Funct Neurosurg (Suppl) 1996; 66: 121–128
7. Kida Y, Kobayashi T, Tanaka T, Oyama H, Iwakoshi T. Stereotactic radiosurgery of intracranial meningiomas. No Shinkei Geka 1994; 22: 621–6
8. Kida Y, Kobayashi T, Tanaka T, Oyama H, Niwa M, Maesawa S. Radiosurgery of cavernous sinus meningiomas with gamma-knife. No Shinkei Geka 1996; 24: 529–33
9. Kondziolka D, Lunsford LD, Coffey RJ, Flickinger JC. Stereotactic radiosurgery of meningiomas. J Neurosurg 1991; 74: 522–559
10. Kondziolka D, Lunsford LD, Coffey RJ, Flickinger JC. Gamma Knife radiosurgery of meningiomas. Stereotact Funct Neurosurg 1991; 57: 11–21
11. Kondziolka D, Lunsford LD. Radiosurgery of Meningiomas. In: Lunsford LD (ed) Stereotactic Radiosurgery 1992; Philadelphia:WB Saunders; pp 219–230
12. Kondziolka D, Lunsford LD, Coffey RJ, Flickinger JC. Gamma knife radiosurgery of meningiomas. Stereotact Funct Neurosurg 1991; 57: 11–21
13. Liscak R, Vladyka V, Simonova G, Novotny J. Radiosurgical treatment of meningioma with the Leksell gamma knife. Cas Lek Cesk 1995; 134: 534–538
14. Lunsford LD. Contemporary management of meningiomas: radiation therapy as an adjuvant and radiosurgery as an alternative to surgical removal? J Neurosurg 1994; 80: 187–190
15. Nakamura S, Hiyama H, Arai K, Nakaya K, Sato H, Hayashi M,

Kawamata T, Izawa M, Takakura K. The Results of Gamma Knife Radiosurgery for Meningiomas: 4 cases of radiation induced edema. Stereotact Funct Neurosurg (Suppl) 1996; 66: 142–145

16. Nicolato A, Ferraresi P, Foroni R, Pasqualin A, Piovan E, Severi F, Masotto B, Gerosa M. Gamma Knife Radiosurgery in Skull Base Meningiomas: Preliminary Experience of 50 Cases. Stereotact Funct Neurosurg (Suppl) 1996; 66: 112–128

17. Pendl G, Schrottner O, Friehs GM, Feichtinger H. Stereotactic radiosurgery of skull base meningiomas. Stereotact Funct Neurosurg 1995; 64 (Suppl 1): 11–8

18. Steiner L, Lindquist C, Steiner M. Meningiomas and Gamma Knife Radiosurgery. In: Al-Mefty O (ed) Meningiomas 1991; New York: Raven Press; pp 263–272

19. Tanaka T, Kobayashi T, Kida Y. Growth control of cranial base meningiomas by stereotactic radiosurgery with a gamma knife unit. Neurol Med Chir (Tokyo) 1996; 36: 7–10

Malignant Secondary Tumours

1. Coffey RJ, Flickinger JC, Bissonette DJ, Lunsford LD. Radiosurgery for solitary brain metastases using the cobalt-60 Gamma Unit: Methods and results in 24 patients. Int J Radiat Oncol Biol Phys 1991; 20: 1287–1295

2. Dong RH, Gao ZW Hu ZQ, Xu WM, Pan L. The preliminary application of Gamma Knife in the treatment of nasopharyngeal carcinoma. Stereotact Funct Neurosurg (Suppl) 1996; 66: 201–207

3. Ericson K, Kihlström L, Mogard J, Karlsson B, Collins VP, Stone-Elander S: Positron Emission Tomography Using 18F-FDG in Patients with Stereotactically Irradiated Brain Metastasis. Stereotact Funct Neurosurg (Suppl) 1996; 66: 214–224

4. Flickinger JC, Kondziolka D, Lunsford LD, Coffey RF, Goodman ML, Shaw EG, Hudgins WR, Weiner R, Harsh GR, Sneed PK, Larson DA. A multi-institutional experience with stereotactic radiosurgery for solitary brain metastasis. Int J Radiat Oncol Biol Phys 1994; 28: 797–802

5. Flickinger JC, Kondziolka D, Lunsford LD, Coffey RJ, Goodman ML, Shaw EG, Hudgins WR, Weiner R, Harsh GR IV, Sneed PK, Larson DA. A multi-institutional experience with stereotactic radiosurgery for solitary brain metastases. Int J Radiat Oncol Biol Phys 1992; 23: 413–418

6. Fukuoka S, Seo Y, Takanashi M, Takahashi S, Suematsu K, Nakamura J-I. Radiosurgery of brain metastases with the Gamma Knife. Stereotact Funct Neurosurg (Suppl) 1996; 66: 193–200

7. Gerosa MA, Nicolato A, Berlucchi S, Piovan E, Zampieri PG, Pasoli A, Foroni R, Giri MG, Marchini G, Babighian S, et al. Gamma Knife radiosurgery of primary and metastatic malignant brain tumors, a preliminary report. Stereotact Funct Neurosurg 1995; 64 (Suppl 1): 56–66

8. Gerosa M, Nicolato A, Severi F, Ferraresi P, Masotto B, Barone G, Oroni R, Piovan E, Pasoli A, Bricolo A. Gamma Knife Radiosurgical Treatment of Intracranial Metastases: From Local Tumor Control to Increased Survival. Stereotact Funct Neurosurg (Suppl) 1996: 66; 184–192

9. Jokura H, Takahashi K, Kayama T, Yoshimoto T. Gamma Knife radiosurgery of a series of only minimally selected metastatic brain tumours. Acta Neurochir Suppl (Wien) 1994; 62: 77–82

10. Karlsson B, Kihlström L, Lindquist C. Medical controversy: Is Gamma Knife surgery the treatment of choice for cerebral metastases? Forum 1994; 4: 396–402

11. Kida Y, Kobayashi T, Tanaka T, Oyama H, Iwakoshi T. Gamma-radiosurgery of metastatic brain tumors. No Shinkei Geka 1993; 11: 991–997

12. Kida Y, Kobayashi T, Tanaka T. Radiosurgery of the metastatic brain tumors with Gamma Knife. Acta Neurochir (Suppl) 1995; 63: 89–94

13. Kihlström L, Carlsson B, Lindquist C. Gamma knife surgery in brain metastases. In: Lunsford LD (ed) Stereotactic Radiosurgery Update 1992; New York, Amsterdam, London, Tokyo: Elsevier; pp 429–434

14. Kihlström L, Karlsson B, Lindquist C, Norén G, Rähn T. Gamma Knife surgery for cerebral metastasis. Acta Neurochir Suppl (Wien) 1991; 52: 87–89

15. Kihlström L, Karlsson B, Lindquist C. Stereotactic radiosurgery for single and multiple cerebral metastases. Acta Neurochirurg 1993; 122: 158–164

16. Kihlström L, Karlsson B, Lindquist C. Gamma Knife surgery for cerebral metastases. Implications for survival based on 16 years experience. Stereotact Funct Neurosurg 1993; 61 (Suppl 1): 45–50

17. Koike Y, Hosoda H, Ishiwata Y, Sakata K, Hidaka K. Effect of radiosurgery using Leksell gamma unit on metastatic brain tumor. Autopsy case report. Neurol Med Chir 1994; 34: 534–537

18. Lang E, Slater J. Metastatic brain tumors: results of surgical and non-surgical treatment. Surg Clin N Am 1964; 44: 865–872

19. Lindquist C. Gamma knife surgery for recurrent solitary metastasis of a cerebral hypernephroma. Case report. Neurosurgery 1989; 25: 802–804

20. Marchini G, Gerosa M, Piovan E, Pasoli A, Babighian S, Rigotti M, Rossato M, Bonomi L. Gamma Knife Stereotactic Radiosurgery or Uveal Melanoma: Clinical Results after Two Years. Stereotact Funct Neurosurg (Suppl) 1996: 66; 208–213

21. Mogard J, Kihlström L, Ericson K, Karlsson B, Guo WY, Stone-Elander S. Recurrent tumor vs radiation effects after gamma knife radiosurgery of intracerebral metastases: diagnosis with PEET-FDG. Comput Assist Tomogr 1994; 18(2): 177–181

22. Rand RW, Jacques DB, Melbye RW, Copcutt BG, Irwin L. Gamma knife radiosurgery for metastatic brain tumors. Acta Neurochir 1995; 63: 85–88

23. Rand RW, Jacques DB, Melbye RW, Copcutt BG, Irwin L. Gamma knife radiosurgery for metastatic brain tumors. Acta Neurochir Suppl (Wien) 1995; 63: 85–88

24. Somaza S, Kondziolka D, Lunsford DL, Kirkwood J, Flickinger JC. Stereotactic radiosurgery for cerebral metastatic melanoma. J Neurosurg 1993; 79(5): 661–666

25. Tago M, Nakagawa K, Terahara A, Aoki Y, Sasaki Y, Kurita H, Kawamoto S, Murayama S. Gamma Knife Radiosurgery for Brain Stem Metastases: Two Autopsy Cases. Stereotact Funct Neurosurg (Suppl) 1996: 66; 225–230

26. Whang CJ, Kwon Y. Gamma knife radiosurgery for malignant tumors. J Korean Med Sci 1995; 10: 379–87

27. Young RF, Jaques DB, Duma C, Rand RW, Henderson J, Vermeulen SS, Grimm P, Blasko JC, Posewitz A, Copcutt B, Bolles GE, Breeze RE, Pribil SG, Winston K, Johnson SD. Gamma Knife radiosurgery for treatment fo multiple brain metastases. In: Kondziolka D (ed) Radiosurgery 1996; Basel: Karger; pp 92–101

Pituitary Tumours

1. Anniko M, Arndt J, Rähn T, Werner S. Gamma irradiation effects on human pituitary adenoma tissue. An analysis of histology, ultrastructure and hormone sampling in an in vitro model system. Acta Otolaryngol (Stockh) 1982; 93: 485–500

2. Backlund EO, Bergstrand G, Hierton-Laurell U, Rosenborg M, Wajnot A, Werner S. Tumor changes after single dose irradiation by stereotactic radiosurgery in 'nonactive' pituitary adenomas and prolactinomas. INSERM 1979; 12: 199–206

3. Backlund E-O, Ganz JC. Pituitary Adenomas and the Gamma Knife. In: Alexander E III, Loeffler JS, Lunsford DL (eds) Stereotactic Radiosurgery 1993; New York: McGraw-Hill; pp 167–174

4. Backlund EO, Rähn T, Sarby B, de Schryver A, Wennerstrand J. Closed stereotaxic hypophysectomy by means of Co-60 gamma radiation. Acta Radiol Ther Phys Biol 1972; 11: 545–555

5. Degerblad M, Rähn T, Bergstrand G, Thorén M. Long term results of stereotactic radiosurgery to the pituitary gland in Cushing's disease. Acta Endocrinol (Copenh) 1986; 112: 310–314

6. Flickinger JC, Lunsford LD, Deutsch M. Retreatment megavoltage irradiation of suprasellar and pituitary tumors. Int J Radiat Oncol Biol Phys 1989; 17: 171–175

7. Ganz JC, Backlund EO, Thorsen FA. The effects of Gamma Knife surgery of pituitary adenomas on tumor growth and endocrinopathies. Stereotact Funct Neurosurg 1993; 61 Suppl 1: 30–37

8. Marek J, Malik J, Fendrych P. Initial experience of an endocrinologist with the treatment of hypophyseal adenomas with the Leksell gamma knife. Cas Lek Cesk 1995; 134: 543–546

9. Pollock BE, Kondziolka D, Lunsford LD, Flickinger JC. Stereotactic radiosurgery for pituitary adenomas: imaging, visual and endocrine results. Acta Neurochir Suppl (Wien) 1994; 62: 33–38

10. Rähn T, Anniko M, Arndt J. Irradiation of human pituitary adenomas in organ culture. INSERM 1979; 12: 213–218

11. Rähn T, Thorén M, Anniko M. Gamma irradiation effects on human ACTH-producing pituitary tumors in organ culture. Arch Otorhinolaryngol 1983; 238: 209–215

12. Sääf M, Thorén M, Bergstrand CG, Norén G, Rähn T, Tallstedt L, Backlund EO. Treatment of craniopharyngiomas-the stereotactic approach in a ten to twenty-three years' perspective. II. Psychosocial situation and pituitary function. Acta Neurochir (Wien) 1989; 99: 97–103

13. Seo Y, Fukuoka S, Takanashi M, Sasaki T, Suematsu K, Nakamura J. Gamma knife surgery for Cushing's disease. Surg Neurol 1995; 43: 170–175

14. Thorén M, Rähn T, Guo W, Werner S. Stereotactic Radiosurgery with the Cobalt-60 Gamma Unit in the Treatment of Growth Hormone-Producing Pituitary Tumors. Neurosurgery 1991; 29: 663–668

15. Thorén M, Rähn T, Hall K, Backlund EO. Treatment of pituitary-dependent Cushing's syndrome with closed stereotactic radiosurgery by means of Co-60 gamma radiation. Acta Endocrinol (Copenh) 1978; 88: 7–17

16. Thorén M, Rähn T, Hallengren B, et al. Treatment of Cushing's disease in childhood and adolescence by stereotactic pituitary irradiation. Acta Paediatr Scand 1986; 75: 338–395

17. Thoren M, Rahn T, Guo WY, Werner S. Stereotactic radiosurgery with the cobalt-60 gamma unit in the treatment of growth hormone-producing pituitary tumors. Neurosurgery 1991; 29: 663–668

18. Vladyka V, Liscak R, Subrt O, Simonova G, Novotny J. Use of the radiosurgery knife in the treatment of hypophyseal adenomas. Cas Lek Cesk 1995; 134: 539–542

Primary Cerebral Tumours

1. Coffey RJ. Boost Gamma Knife radiosurgery in the treatment of primary glial tumors. Stereotact Funct Neurosurg (Suppl) 1993; 61: 59–64

2. Dempsey PK, Kondziolka D, Lunsford LD, Coffey RJ, Flickinger JC. The role of stereotactic radiosurgery in the treatment of glial tumors. In: Lunsford LD (ed) Stereotactic Radiosurgery Update 1992; New York: Elsevier: pp 407–413

3. Ganz JC, Smievoll AI, Thorsen F. Radiosurgical treatment of gliomas of the diencephalon. Acta Neurochir Suppl (Wien) 1994; 62: 62–66

4. Gerosa MA, Nicolato A, Berlucchi S, Piovan E, Zampieri PG, Pasoli A, Foroni R, Giri MG, Marchini G, Babighian S, et al. Gamma Knife

radiosurgery of primary and metastatic malignant brain tumors, a preliminary report. Stereotact Funct Neurosurg 1995; 64 (Suppl 1): 56–66

5. Grabb PA, Lunsford LD, Albright AL, Kondziolka D, Flickinger JC. Stereotactic radiosurgery for glial neoplasms of childhood. Neurosurgery 1996; 38: 696–701

6. Inoue H, Kohga H, Zama A, Hirato M, Shibazaki T, Andou Y, Ohye C. Pathobiology of Cerebral Gliomas in Children and the Role of Radiosurgery. Stereotact Funct Neurosurg (Suppl) 1996: 66; 278–287

7. Inoue HK, Kunimine H, Zama A, Ono N, Nakamura M. Clinical pathology of primitive gliomas in the cerebrum. Acta Neurochir (Wien) 1986; 81: 94–99

8. Inoue HK, Nakamura M, Ono N, Kawashima Y, Hirato M, Ohye C. Long-term clinical effects of radiation therapy for primitive gliomas and medulloblastomas: A role of radiosurgery. Stereotact Funct Neurosurg 1993; 61 (Suppl 1): 51–58

9. Kihlström L, Lindquist C, Lindquist M, Karlsson B. Stereotactic radiosurgery for tectal low-grade gliomas. Acta Neurochir Suppl (Wien) 1994; 62: 55–57

10. Kondziolka D, Lunsford LD, Claassen D, Pandalai S, Maitz AH, Flickinger JC. Radiobiology of radiosurgery, Part II: The rat C6 glioma model. Neurosurgery 1992; 31: 280–288

11. Larson DA, Bova F, Eisert D, Kline R, Loeffler J, Lutz W, Metha M, Palta J, Schewe K, Schultz C, Shaw E, Wilson F. Current radiosurgery practice: results of an astro survey. Int J Radiat Oncol Biol Phys 1993; 28: 523–526

12. Lim YJ, Leem W: Two Cases of Gamma Knife Radiosurgery for Low Grade Optic Chiasm Glioma. Stereotact Funct Neurosurg (Suppl) 1996: 66; 174–184

13. Loeffler JS, Alexander E, Shea WM, Wen PY, Fine HA, Kooy HM, Black P. Radiosurgery as part of the initial management of patients with malignant gliomas. J Clin Oncol 1992; 10: 1279–1385

14. Loeffler JS, Shrieve DC, Alexander III E. Radiosurgery for glioblastoma multiforme: the importance of selection criteria. Int J Radiat Oncol Biol Phys 1994; 30: 731–733

15. Manera L, Régis J, Chinot O, Porcheron D, Levrier O, Farnarier P, Peragut JC. Pineal Region Tumors: The Role of Stereotactic Radiosurgery? Stereotact Funct Neurosurg (Suppl) 1996: 66; 164–173

16. Marsa GW, Goffinet DR, Rubinstein LJ, et al. Megavoltage irradiation in the treatment of gliomas of the brain and spinal cord. Cancer 1975; 36: 1681–1689

17. Pollack IF, Claassen D, Al-Shboul Q, Janosky JE, Deutsch M. Low-grade gliomas of the cerebral hemispheres in children: an analysis of 71 cases. J Neurosurg 1995; 82: 536–547

18. Whang CJ, Kwon Y. Gamma knife radiosurgery for malignant tumors. J Korean Med Sci 1995; 10: 379–387

19. Yoshino E, Ohmori Y, Imahori Y, Higuchi T, Furuya S, Naruse S, Mori T,

Suzuki K, Yamaki T, Ueda S, Tsuzuki T, Takai S: Irradiation Effects on the Metabolism of Metastatic Brain Tumors: Analysis by PET and 1H-MRS Studies. Stereotact Funct Neurosurg (Suppl) 1996; 66: 240–260

Functional Disease

1. Bingley T, Leksell L, Meyerson BA, Rylander G. Long time results of stereotactic anterior capsulotomy in chronic obsessive-compulsive neurosis. In: Sweet WH, Obrador S, Martin-Rodriguez JG (eds) Neurological Treatment in Psychiatry, Pain and Epilepsy 1977; Baltimore: University Park Press; pp 287–299

2. Boethius J, Forster DMD, Leksell L, Meyerson BA, Steiner L. Medial thalamotomy in cancer pain. Acta Neurochir (Wien) 1973; 29: 263

3. Burcheil KJ. Thalamotomy for movement disorders. In: Gildenberg PL (ed) Neurosurgery Clinics of North America, Functional Neurosurgery 1995; Philadelphia: WB Saunders; pp 55–72

4. Elomaa E. Focal irradiation of the brain: An alternative to temporal lobe resection in intractable focal epilepsy. Med Hypotheses 1980; 6: 501–503

5. Forster DMC, Leksell L, Meyerson BA, Steiner L. Gammathalamotomy in intractable pain. In: Janzen RW, Keidel WD, Herz A, Steichele C (eds) Pain. Basic Principles, Pharamacology, Therapy 1972; Stuttgart: George Thieme Verlag; pp 194–198

6. Friedman JH, Epstein M, Sanes JN, Lieberman P, Cullen K, Lindquist C, Daamen M. Gamma knife pallidotomy in advanced Parkinson's disease. Ann Neurol 1996; 39: 535–538

7. Friehs GM, Norén G, Ohye C, Duma CM, Marks R, Plombon J, Young RF. Lesion Size Following Gamma Knife Treatment for Functional Disorders. Stereotact Funct Neurosurg (Suppl) 1996: 66; 320–328

8. Friehs GM, Ojakangans CL, Pachatz P, Schröttner O, Ott E, Pendl G. Thalamotomy and Caudatotomy with the Gamma Knife as a Treatment for Parkinsonism with a Comment on Lesion Sizes. Stereotact Funct Neurosurg 1995; 64 (Suppl 1): 209–221

9. Heikkinen ER, Konnov B, Melinikov L, Yalynych N, Zubkov Y, Garmashov Y, Pak VA. Relief of epilepsy by radiosurgery of cerebral arteriovenous malformations. Stereotact Funct Neurosurg 1989; 53: 157–166

10. Hellstrand E, Abraham-Fuchs K, Jernberg B, Kihlström L, Knutsson E, Lindquist C, Schneider S, Wirth A. MEG localization of interictal epileptic focal activity and concomitant stereotactic radiosurgery. A non-invasive approach for patients with focal epilepsy. Physiol Meas 1993; 14: 131–136

11. Hirai T, Miyazaki M, Nakajima H, Shibazaki T, Ohye C. The correlation between tremor characteristics and predicted volume of effective lesions in stereotaxic Vim-thalamotomy. Brain 1983; 106: 1001–1018

12. Hirato M, Ohye C, Shibazaki T, Nakamura M, Inoue HK, Andou Y.

Gamma Knife Thalamotomy for Treatment of Functional Disorders. Stereotact Funct Neurosurg 1995; 64 (Suppl 1): 165–171

13. Kelly PJ. Comment on Ostuki, et al: 'Stereotactic gamma-thalamotomy with a computerized brain atlas': technical case report. Neurosurg 1994; 35: 768

14. Kihlström L, Guo W, Lindquist C, Mindus P. Radiobiology of radiosurgery for retractory anxiety disorders. Neurosurgery 1995; 36: 294–302

15. Kondziolka D, Flickinger JC, Lunsford LD, Habeck M: Trigeminal Neuralgia Radiosurgery: The University of Pittsburgh Experience. Stereotact Funct Neurosurg (Suppl) 1996: 66; 343–349

16. Kondziolka D, Lunsford LD, Flickinger JC, Young RF, Vermeulen S, Duma CM, Jacques DB, Rand RW, Regis J, Peragut JC, Manera L, Epstein MH, Lindquist C. Stereotactic radiosurgery for trigeminal neuralgia: a multiinstitutional study using the gamma unit. J Neurosurg 1996; 84: 940–945

17. Leksell L. Cerebral radiosurgery. 1. Gammathalamotomy in two cases of intractable pain. Acta Chir Scand 1968; 134: 585–595

18. Leksell L. Stereotaxic Radiosurgery in trigeminal neuralgia. Acta Chir Scand 1971; 37: 311–314

19. Leksell L. Cerebral radiosurgery I. Gammathalamotomy in two cases of intractable pain. Acta Chir Scand 1968; 134: 585–595

20. Leksell L. Stereotaxic radiosurgery in trigeminal neuralgia. Acta Chir Scand 1971; 137: 311–314

21. Leksell L, Backlund EO. Stereotactic gamma capsulotomy. In: Hitchcock ER, Ballantine HT Jr, Meyerson BA (eds) Modern Concepts in Psychiatric Surgery 1979; Amsterdam: Elsevier/North Holland Biomedical Press; pp 213–216

22. Leksell L, Meyersom BA, Forster DMC. Radiosurgical thalamotomy for intractable pain. Confin Neurol 1972; 34: 264

23. Lindquist C, Kihlström L, Hellstrand E. Functional neurosurgery – A future for the Gamma Knife. Stereotact Funct Neurosurg 1991; 57: 72–81

24. Mindus P, Bergström K, Levander SE, Norén G, Hindmarch T, Thuomas KA. Magnetic resonance images related to clinical outcome after psychosurgical intervention in severe anxiety disorder. J Neurol Neurosurg Psychiat 1987; 50: 1288–1293

25. Mindus P, Meyerson BA. Anterior capsulotomy for intractable anxiety disorders. In: Schmidek HH, Sweet WH (eds) Operative Neurosurgical Techniques 1995; Philadelphia: WB Saunders Company; pp 1443–1455

26. Mindus P, Rasmussen SA, Lindquist C. Neurosurgical treatment for refractory obsessive-compulsive disorder. Implications for understanding frontal lobe function. J Neuropsychiatry 1995; 6: 467–477

27. Mindus P, Rauch SL, Nyman H, Baer L, Edman G, Jenike MA. Capsulotomy and cingulotomy as treatments for malignant obsessive compulsive disorder. An update. In: Hollander E, Zohar J, Marazzati D,

Olivier B (eds) Current Insights in Obsessive Compulsive Disorder 1994; New York: J. Wiley; pp 245–276

28. Ohye C, Shibazaki T, Hirato M, Inoue H, Andou Y: Gamma Thalamotomy for Parkinsonian and Other Kinds of Tremor. Stereotact Funct Neurosurg (Suppl) 1996: 66; 333–342

29. Otsuki T, Jokura H, Takahashi K, Ishikawa S, Yoshimoto T, Kimura M, Yoshida R, Miyazawa T. Stereotactic gamma-thalamotomy with a computerized brain atlas. Technical case report. Neurosurgery 1994; 35: 764–768

30. Pan L, Dai JZ, Wang B-J, Xu WM, Zhou L-F, Chen X-R. Stereotactic Gamma Thalamotomy for the Treatment of Parkinsonism. Stereotact Funct Neurosurg (Suppl) 1996: 66; 329–332

31. Pendl G, Schrottner O, Friehs GM, Legat J, Leber K, Mokry M, Papaefthymiou G, Langmann G. Radiosurgery with the first Austrian cobalt-60 Gamma-unit. A one year experience. Acta Neurochir (Wien) 1994; 127: 170–179

32. Rand RW, Jacques DB, Melbye RW, Copcutt BG, Fisher MR, Levenick MN. Gamma Knife thalamotomy and pallidotomy in patients with movement disorders: preliminary results. Stereotact Funct Neurosurg 1993; 61 Suppl 1: 65–92

33. Rand RW, Jacques DB, Melbye RW, Copcutt BG, Levenick MN, Fisher MR. Leksell Gamma Knife treatment of tic douloureux. Stereotact Funct Neurosurg 1993; 61 (Suppl 1): 93–102

34. Régis J, Kerkerian-Legoff L, Rey M, Vial M, Porcheron D, Nieoullon A, Peragut JC: First Biochemical Evidence of Gamma Knife Functional Differential Effects. Stereotact Funct Neurosurg 1996; (Suppl) 66: 29–38

35. Régis J, Manera L, Dufour H, Porcheron D, Sedan R, Peragut JC. Effect of the Gamma Knife on Trigeminal Neuralgia. Stereotact Funct Neurosurg 1995; 64 (Suppl 1): 182–192

36. Régis J, Peragut JC, Rey M, Samson Y, Levrier O, Porcheron D, Régis H, Sedan R. First Selective Amygdalohippocampal Radiosurgery for 'Mesial Temporal Lobe Epilepsy'. Stereotact Funct Neurosurg 1995; 64 (Suppl 1): 193–201

37. Rinaldi PC, Young RF, Albe-Fessard D, et al. Spontaneous neuronal hyperactivity in the medial and intralaminar thalamic nuclei of patients with deafferentation pain. J Neurosurg 1991; 74: 415–421

38. Rylander G. Stereotactic radiosurgery in anxiety and obsessive-compulsive neurosis. In: Hitchcock ER, Ballantine HTJ, Meyerson BA (eds) Modern Concepts in Psychiatric Surgery 1979; Amsterdam: Elsevier: pp 235–240

39. Steiner L, Forster DMC, Leksell L, Meyerson BA, Boethius J. Gammathalamotomy in intractable pain. Acta Neurochir (Wien) 1980; 52: 173–184

40. Whang CJ, Chang JK. Short Term Follow-Up of Stereotactic Gamma Knife Radiosurgery in Epilepsy. Stereotact Funct Neurosurg 1995; 64 (Suppl 1): 202–208

41. Whang CJ, Kwon Y. Long-term Follow-up of Stereotactic Gamma Knife Radiosurgery in Epilepsy. Stereotact Funct Neurosurg (Suppl) 1996: 66; 349–356

42. Young RF. Intracranial procedures for pain management. In: Apuzzo MLG (ed) Brain Surgery: Complication Avoidance and Management 1993; New York: Churchill Livingstone; pp 1497–1508

43. Young RF. Stereotactic surgical ablation for pain relief. In: Rengachary SS, Wilkins RH (eds) Neurosurgical operative atlas; Baltimore: Williams and Wilkins; pp 177–188

44. Young RF. Gamma Knife treatment of Parkinson's disease. Northwest Neuroscience Institute Journal 1996; 3: 1–2

45. Young RF, Jacques DS, Rand RW, Copcutt BC, Vermeulen SS, Posewitz AE. Technique of stereotactic medial thalamotomy with the Leksell Gamma Knife for treatment of chronic pain. Neurol Res 1995; 17: 59–65

46. Young RF, Jacques DS, Rand RW, et al. Medial thalamotomy with the Leksell Gamma Knife for treatment of chronic pain. Acta Neurochir (Suppl) 1994; 62: 105–110

47. Young RF, Jacques DS, Rand RW, et al. Technique of stereotactic medial thalamotomy with the Leksell Gamma Knife for treatment of chronic pain. Neurol Res 1995; 17: 59–65

48. Young RF, Rand DSJ, Copcutt BR. Medial Thalamotomy with the Leksell Gamma Knife for treatment of chronic pain. Acta Neurochir Suppl (Wien) 1994; 62: 105–110

49. Young RF, Vermeulen SS, Grimm P, Posewitz AE, Jacques DB, Rand RW, Copcutt BG. Gamma Knife Thalamotomy for the Treatment of Persistent Pain. Stereotact Funct Neurosurg 1995; 64 (Suppl 1): 172–181

50. Young RF, Vermeulen SS, Posewitz A, Grimm P, Blasko J, Jaques DB, Rand RW, Copcutt B. Functional neurosurgery with the Leksell Gamma Knife. In: Kondziolka D (ed) Radiosurgery 1995 1996; Basel: Karger; pp 218–228

51. Young RF, Vermeulen SS, Grimm P, Posewitz A. Electrophysiological Target Localization is not Required for the Treatment of Functional Disorders. Stereotact Funct Neurosurg (Suppl) 1996: 66; 309–319

52. Young RF, Jacques DS, Rand RW, Copcutt BC, Vermeulen SS, Posewitz AE. Technique of stereotactic medial thalamotomy with the Leksell Gamma Knife for treatment of chronic pain. Neurol Res 1995; 17: 59–65

53. Young RF, Jacques DS, Rand RW, Copcutt BR. Medial thalamotomy with the Leksell Gamma Knife for treatment of chronic pain. Acta Neurochir Suppl (Wien) 1994; 62: 105–110

Subject Index

Springer Neurosurgery

C. Lindquist, D. Kondziolka,
J. S. Loeffler (eds.)

Advances in Radiosurgery

Proceedings of the 1st Congress of the International
Stereotactic Radiosurgery Society, Stockholm 1993

1994. 71 figures. VIII, 124 pages.
Cloth DM 165,–, öS 1155,–
Reduced price for subscribers to "Acta Neurochirurgica":
Cloth DM 148,50, öS 1039,50
ISBN 3-211-82612-2
Acta Neurochirurgica, Supplement 62

Radiosurgery is a rapidly developing form of minimally
invasive neurosurgery. Selected papers from the first meeting
of the International Stereotactic Radiosurgery Society in
Stockholm, June 1993, reflect current multidisciplinary
approaches to difficult intracranial neurosurgical problems.
Neurosurgeons, radiotherapists, oncologists, radiobiologists,
physicists and representatives of several other clinical disci-
plines inform about the state-of-the-art of radiosurgical treatment of a multitude of intracranial problems such as arteriovenous malformations, pituitary and pineal tumors, vestibular
schwannomas as well as metastatic brain tumors and gliomas.

Springer Wien New York

Sachsenplatz 4-6, P.O.Box 89, A-1201 Wien, Fax +43-1-330 24 26,
e-mail: order@springer.co.at, Internet: http://www.springer.co.at
New York, NY 10010, 175 Fifth Avenue • Heidelberger Platz 3, D-14197 Berlin
Tokyo 113, 3-13, Hongo 3-chome, Bunkyo-ku

Springer News Neurosurgery

C. B. Ostertag, D. G. T. Thomas,
A. Bosch, B. Linderoth,
G. Broggi (eds.)

Advances in Stereotactic and Functional Neurosurgery 12

1997. 59 figures. X, 144 pages.
Cloth DM 120,–, öS 840,–
Reduced price for subsribers to "Acta Neurochirurgica":
Cloth DM 108,–, öS 756,–
ISBN 3-211-82978-4
Acta Neurochirurgica, Supplement 68

The proceedings of the 12th Congress of the European Society for Stereotactic and Functional Neurosurgery in Milan contain selected contributions on surgery of Parkinson's disease, pain, psychosurgery, epilepsy, frameless stereotaxy, functional imaging, gene therapy and radiosurgery. The selection reflects the current status and progress in the field. The book is an update of the current methods and controversies and should serve those specialized in the field as a source of information and judgement. The foreword by Jens Haase unifies the views of both stereotactic and general neurosurgeons with great enthusiasm.

Springer Wien New York

Sachsenplatz 4-6, P.O.Box 89, A-1201 Wien, Fax +43-1-330 24 26.
e-mail: order@springer.co.at, Internet: http://www.springer.co.at
New York, NY 10010, 175 Fifth Avenue • Heidelberger Platz 3, D-14197 Berlin
Tokyo 113, 3-13, Hongo 3-chome, Bunkyo-ku

Springer-Verlag
and the Environment

WE AT SPRINGER-VERLAG FIRMLY BELIEVE THAT AN international science publisher has a special obligation to the environment, and our corporate policies consistently reflect this conviction.

WE ALSO EXPECT OUR BUSINESS PARTNERS – PRINTERS, paper mills, packaging manufacturers, etc. – to commit themselves to using environmentally friendly materials and production processes.

THE PAPER IN THIS BOOK IS MADE FROM NO-CHLORINE pulp and is acid free, in conformance with international standards for paper permanency.